Naturally . . . South Texas

Corrie Herring Hooks
Number Forty-eight

Naturally . . . South Texas

Nature Notes from the Coastal Bend

ROLAND H. WAUER

DRAWINGS BY MIMI HOPPE WOLF

University of Texas Press, *Austin*

Printed in the United States of America

First edition, 2001

Requests for permission to reproduce material from this work should be sent to Permissions, University of Texas Press, P.O. Box 7819, Austin, TX 78713-7819.

The paper used in this book meets the minimum requirements of ANSI/NISO Z39.48-1992 (R1997) (Permanence of Paper).

Library of Congress Cataloging-in-Publication Data

Wauer, Roland H.

Naturally—South Texas / Roland H. Wauer.—1st ed.

p. cm.

Includes index.

ISBN 0-292-79144-5 (cloth : alk. paper)—ISBN 0-292-79139-9 (pbk : alk. paper)

1. Natural history—Texas, South. 2. Seasons—Texas, South. I. Title.

QH105.T4 W39 2001

508.764'1—dc21 00-050312

This book of South Texas nature notes is dedicated to all the members of the Golden Crescent Nature Club.

Contents

Introduction

It seems to me that I have been writing nature notes for as long as I can remember. Even before high school I wrote short observations about tropical fish that I raised, various birds, snakes, and lizards that I found in a nearby field, an injured American Kestrel that I nursed back to health, and almost anything else that caught my attention. All of those notes have been lost along the way.

Later, when my "fork in the road" led me into the National Park Service, where I worked for thirty-two years before retiring in 1989, I found myself writing nature notes on the various parks in which I worked: Crater Lake, Oregon Caves, Death Valley, Pinnacles, Yosemite, Zion, Big Bend, Bandelier, Great Smoky Mountains, and Virgin Islands.

When the Park Service began to emphasize interpretive outreach, I also began writing nature notes—maybe they should have been called *environmental theses*—for several local newspapers. It was a good way to inform readers about our parks and environmental concerns, to help people better understand the natural world that we all must depend upon for our long-term subsistence. I felt it was a significant contribution. Perhaps African conservationist Baba Dioum expressed that reasoning best: "For in the end, we will conserve only what we love. We will love only what we understand, and we will understand only what we are taught."

After I retired from the National Park Service, Betty and I moved to Victoria, primarily so that she could be closer to her four sons and their families. South Texas also fit well with my agenda of enjoying the outdoors in what undoubtedly is the best birding area anywhere in the United States. However, it was almost immediately apparent to me that the local newspaper, the *Victoria Advocate*, did not give adequate attention to the natural world around us. Although news of dramatic natural events was well covered, day-to-day happenings in nature, those things that I happened to be most interested in, were often ignored.

It was almost a year later when visiting with Bill Farnsworth, president of the Golden Crescent Nature Club at the time, about my concern that he suggested that we talk with *Victoria Advocate* publisher John Roberts about a weekly nature column. John was immediately interested, and my first note—"Neotropical Migrants Take Wing over Golden Crescent"—appeared on Sunday, April 24, 1994. Although only a handful of notes were written during that first summer due to my extensive travel schedule, my column has appeared every Sunday since then. A few of the early notes were written by other club members, namely Ken Bruns, Joe Crisp, Mark Elwonger, Bill and Judie Farnsworth, Elaine Giessel, and Linda Valdez. I thank them one and all! I also want to thank Sara Hendricks, Liz Dechert, and Chari Prenzler, of the "Life Style" section of the *Victoria Advocate*, for their editorial assistance.

The Golden Crescent region of South Texas, the principal area covered by the *Advocate*, includes fifteen counties that extend along the central Gulf Coast from Matagorda County south through Aransas County and west through Karnes, Gonzales, and Fayette Counties. Readership is estimated at 40,000. This region of Texas is often called the "Crossroads" for numerous reasons. Biologically it encompasses four rather distinct ecosystems, all within a mile circle of Victoria: the northeastern edge of the South Texas Plains, the southern edges of the Post Oak Savannah and Blackland Prairie, and the heart of the Gulf Prairie and Marshes.

A closer look at the four zones further explains why the Crossroads contains such an environmental diversity. Considering the bird life only, the nesting ranges of the Ruby-throated Hummingbird, Red-bellied Woodpecker, Great Crested Flycatcher, American Crow, Blue Jay, and Carolina Chickadee, all common nesting birds in the East, extend only to the southern edge of the Post Oak Savannah and Blackland Prairie. Those edges are generally marked by the San Antonio River.

On the other hand, several South Texas Plains breeding birds are limited by the forested elements of Victoria and northern Calhoun Counties: Buff-bellied Hummingbird, Golden-fronted Woodpecker, Brown-crested Flycatcher, Long-billed Thrasher, and Olive Sparrow. All of these birds are common in South Texas and northeastern Mexico.

Western species that reach our region include the Harris's Hawk, Ladder-backed Woodpecker, Cactus Wren, Curve-billed Thrasher, Pyrrhuloxia, and Lesser Goldfinch. And the Gulf Prairie and Marshes zone has only one nesting land bird that is unique, the White-tailed Hawk. It is common throughout the South Texas Plains. Similar things could also be said about all the other animals as well as numerous plant species. The Golden Crescent of South Texas represents a true crossroads of biological affinities.

The idea of incorporating a series of my nature notes into a book is based upon a similar book by Henry Wolf, Jr., another *Advocate* columnist who has written "Henry's Journals" five days a week since 1979. When I approached Shannon Davies, former acquisition editor at the University of Texas Press, about the idea, she was immediately interested and initiated a written agreement soon afterward.

The next step was to select a series of my nature notes from those already published in the *Victoria Advocate*. My intention was to provide sufficient topics to cover a full year, chronologically, so that anyone interested in any one month or period of time could learn about what happens in the Golden Crescent region of South Texas during that time frame. Nature notes like these also provide a record of the changes that occur in our region over time.

I hope that the material that follows helps you to better understand and therefore appreciate and help protect our fragile natural world. And finally, a quote from John Harsen Rhoades:

> Do more than exist—live.
> Do more than touch—feel.
> Do more than look—observe.
> Do more than listen—understand.
> Do more than talk—say something!

Enjoy!
—Ro Wauer

Natural History Calendar

JANUARY

Coldest month

Cardinals begin to sing

Bald Eagles nest

Wintering butterflies appear at early flowering shrubs

Live Oak Caterpillars appear in the oaks

FEBRUARY

Spring is in the air

First Purple Martins return

Redbud trees flower

First neotropical migrants appear

Spanish Daggers bloom

Crane Flies appear

Huisache trees flower in mass

MARCH

Days usually are warm and sunny

Ruby-throated Hummingbirds appear

Cliff Swallows return to nest sites

Scissor-tailed Flycatchers return

Spring solstice

Robin and Waxwing flocks pass through

Striped Skunks start hunting mates

Neotropical migrants increase

Bald Eagles head north
Watch out for chiggers

APRIL

Wildflowers peak
Whooping Cranes leave Aransas, heading north
Neotropical migrants continue to increase
Chimney Swifts return to our neighborhoods
First fireflies are active during evening hours
Mesquite leaves green up

MAY

Northbound migrants peak the first few days
Retamas produce bright yellow flowers
Barred Owl youngsters are out and about
Annual Neotropical Migrant Count
Yellow-billed Cuckoo calls are commonplace
Camel Crickets can be abundant

JUNE

Copperheads are again out and about
Bald Cypress trees are in full summer dress
Painted Buntings are commonplace
Crested Caracara young are in training
Longest day of the year
Wood Storks appear in our wetlands

JULY

Thunderstorms can be expected
Aransas and Victoria July 4 butterfly counts
Early southbound shorebirds appear
Wild grapes ripen and are eaten by wildlife and humans
Daddy Longlegs come out of hiding

AUGUST

Hottest month
Garden spiderwebs become numerous

Praying Mantises increase
Mississippi Kites appear over our towns
Fairy rings arise after each storm
Southbound migrants increase
Tropical storms are possible

SEPTEMBER

Wettest month
Field Crickets can be abundant
Southbound Ruby-throated Hummingbirds are everywhere
Bald Eagles return to their nesting grounds
Last of the Ruby-throated Hummingbirds move south
Eastern Phoebes return for the winter months
Significant hawk migration gets underway

OCTOBER

Peregrine Falcons return
Leafcutter Ants are especially active
Monarch Butterflies are migrating south en masse
Sandhill Cranes begin to arrive
Flocks of American White Pelicans appear in the skies
Red Admiral Butterflies appear

NOVEMBER

Cooling trend noticeable
Numbers of Snow and White-fronted Geese appear
Early Whooping Cranes arrive at Aransas
"Butter-butts" return for the winter
Fall color appears on a few oaks and other broadleaf trees
Red berries appear on several shrubs

DECEMBER

Holiday spirit prevails
Winter solstice
Christmas Bird Counts dominate the last days of the month

January

Bird Feeding Is the Perfect Way to Enjoy Nature

JANUARY 1, 1995

How many of you feed birds? According to a report by the U.S. Fish and Wildlife Service, more than 86 million Americans regularly put out feed for birds. That's one out of every three people in the United States. And those 86 million Americans spend more than one billion dollars on birdseed annually.

An additional amount is spent on feeders, binoculars, and such. The hobby of feeding birds is increasing for a very good reason—it is the best possible way to enjoy the outdoors from the comfort of our homes.

The easiest group of birds to attract to a home feeder is the seed-eaters. Currently a dozen or more species around our yards eat seeds. A few years ago, birdseed was pretty well limited to cracked or whole corn, but now there are a variety of options. And there is considerable information about which seed is best for what species.

The most preferred birdseed available today, without doubt, is black-oil sunflower. Striped sunflower seeds are larger and have thicker seed coats that are difficult to handle and too tough for smaller birds. Carolina Chickadees, Tufted Titmice, Northern Cardinals, sparrows, and finches all prefer black-oil sunflower seeds.

The standard birdseed, common on shelves in all the stores, contains a blend of sunflower, milo, millet, oats, wheat, flax, and buckwheat seeds. Many birds will kick out some seeds to get to the prized ones. This results in lots of unused seeds and greater expense; the more expensive black-oil sunflower seeds are cheaper in the long run.

Birdseed feeder with two Northern Cardinals, a Carolina Chickadee, and a Tufted Titmouse.

Niger or thistle is preferred by finches, especially our American Goldfinches and Pine Siskins. However, because these seeds are so small, special feeders are necessary. Also, cracked corn is still popular for Blue Jays, Mourning and Inca Doves, and Northern Bobwhites.

Another option in winter is suet, favored by woodpeckers but also eaten by chickadees, titmice, Carolina and House Wrens, and cardinals. Suet is strictly a wintertime food; it turns rancid when temperatures exceed 70 degrees. Birds prefer plain, inexpensive beef suet over commercial suet cakes. Suet can be placed in wire baskets, wired to trees, or pressed into holes drilled in small logs hung from trees. Wire baskets are recommended because they are less likely to get oil on the bird's plumage.

You may want to make your own suet food. Melt suet for shortening and blend it with yellow cornmeal, two eggs, and water; place in a 12 × 15-inch pan, put in the oven, and bake at 450 degrees for 20 to 25 minutes. Cornbread on a feeder tray will also be a big hit with most of your feeder birds. Plus, I fix a mixture of peanut butter and cornmeal that can then be packed in the holes in feeder logs.

Nectar-feeders normally are few and far between in winter, but this year several hummingbirds have remained with us. Rufous and Buff-bellied Hummingbirds are most numerous, and Christmas Bird Counters also recorded a couple lingering Ruby-throats as well. Although hummingbird diets normally consist of more than 50% insects, they will continue to use your feeders if maintained all winter.

Feeder maintenance is extremely important year-round. Seed-feeders should be cleaned and scrubbed with soap and water, dipped or washed with a solution of bleach (1 to 9), and dried thoroughly each season. Also rake up the seed hulls regularly; decomposed hulls will kill the lawn and could spread disease to your birds. Feeding birds is great fun and a way to attract birds to your yard; my yard list includes 135 species to date.

The Earth Needs Everyone's Resolve

JANUARY 4, 1998

By now, most everyone who makes New Year's resolutions has already done so. But I suspect that many of you have already lost your resolve. It happens every year. I suppose that is part of our culture. On the other hand, a few folks, for whatever reason, may not have decided on a really good resolution yet. Or, because they already have given up on the first one, they may need to start over again.

So, I have a suggestion: Why not make a resolution to do something positive for the environment? Something that would truly be worthwhile and, in a large sense, affect our outdoor environment, which, in the long term, can lead to our own greater health and happiness.

How about starting with a resolution to keep our roadways and fields cleaner? Not dumping garbage and other unwanted things along the roads, in our streams, and under bridges and such, not only will save our tax dollars but also will help control pests. I realize that only a few readers are such slobs who actually go so far as to dump garbage and larger unwanted stuff, but even smaller things like gum wrappers and empty cans and cups accumulate. Even those small things attract pests such as cockroaches and mosquitoes that may be hazardous to our health.

For cat and dog lovers, how about a resolution to keep your pet indoors or controlled so that it does not run wild? Not only do dogs annoy your neighbors by barking and leaving scat on their lawns and gardens, but many dogs, especially those that roam in packs (even within the neighborhood), catch and kill native birds and small mammals, such as cottontails, that many of us enjoy. Cats are especially aggressive, and even a "house cat" that is left outdoors during the daytime will undoubtedly revert back to its natural behavior and kill songbirds. No matter how "domesticated" your loving cat may be, your tabby is a marvelous creature that has evolved into a killing machine. Anyone who lets his or her house cat run free has little respect for the natural environment.

Another possible New Year's resolution, especially for retirees, might be to give some of your time to a good cause. There are lots of those around. For those readers with special interests in nature and/or history, how about volunteering at an area park or refuge? It doesn't take an academic degree to help on maintenance or clerical projects. Most of our local parks and refuges, as well as our Texas Zoo and gardens, can use your assistance. Call and ask: it's that simple.

You don't want to leave home, but you would like to help protect the environment? Then how about writing a letter each week? It would take only a little of your time and a 32-cent stamp. Select an issue you care about, let the president and your congressional representatives know how you feel on national issues, and let your governor and state representatives know how you feel about state issues. If you think that your letters won't count, you are wrong. Writing letters is one thing that the average citizen can do to make a difference.

And finally, how about a resolution to do at least one simple, money-saving thing at home? Turn off the lights when they are not in use. It's such a simple idea that many folks just don't get to it. Turning off the lights that are not needed will not only save your own dollars but also lessen the demand for all the energy production that requires coal-burning, water, or nuclear power.

See how easy it is to help protect the environment?

Northern Harriers Winter in South Texas

JANUARY 7, 1996

Northern Harriers, long-winged hawks with obvious white rumps that winter in our fields and pastures, are back with us for the winter months. This wintering raptor is also known as Marsh Hawk, Hen Harrier, Mouse Hawk, or White-rumped Hawk. Whatever name you call it, the Northern Harrier deserves special recognition for a number of reasons.

First, it is North America's only representative of ten harrier species found worldwide. Each is slender, medium sized, long legged, and long tailed. All ten are of the genus *Circus*, a Greek term referring to their circling flight. The species name for our Northern Harrier is *cyaneus*, Latin for blue, referring to the adult male's slaty color.

Second, unlike most hawks, adult males and females possess very different plumage. The contrasting black-and-white male is very different from the buff-colored, striped female. Juveniles look very much like the female. All, however, possess the noticeable white rump-patch.

Third, harriers utilize a very different method of hunting prey. That method has given them their name "harrier," meaning to cruise back and forth low over the open ground in search of prey. While slowly flying back and forth, often searching one area at a time, their wings are held in a shallow V pattern. And instead of depending principally on sight, like all the other hawks, harriers utilize audio clues in their search. They are able to zero in on sounds, such as rodent squeaks, and then immediately double back and pounce on their prey.

Their audio-location ability is performed by an amazing system of triangulation, similar to that used by owls. A close look at a Northern Harrier will reveal sound-reflecting disks, special facial characteristics that are missing from other hawks.

In addition, studies of nesting harriers have revealed that 25% of nesting females, usually subadults, are polygamists, several females mating with one male.

Once incubation begins, the male harrier rarely visits the nest, leaving those chores to the hens. He does provide his fair share of food for the nestlings, by transferring prey in flight to the female, who then sneaks back to the nest after several false landings to confuse any watching predators.

Our wintering Northern Harriers leave South Texas in spring to return to their nesting grounds—from the Texas Panhandle northward across the northern half of the United States—but they are with us until then. Watching these curious raptors hunt rodents and birds in our fields cannot help but increase our appreciation of our diverse natural environment.

Robins Still Bobbin' Along in Golden Crescent

JANUARY 8, 1995

Most winters, American Robins arrive in the Golden Crescent in late fall, consume all the berries available, and then continue south by the end of the year. This year, however, with our warm weather and less competition from some of the other northern species, such as Cedar Waxwings, robins are still with us.

The 1994 Victoria Christmas Bird Count tallied a total of 3,275 robins, compared with fewer than 500 the year before. The "robin-red" breasts, flying across the sky in their scattered formation, were a wonderful sight for the Christmas counters and anyone else lifting their eyes to the heavens.

Robins have long been one of my favorite birds, and for many people they are the harbinger of spring, the coming of the new season. In South

Texas, although they normally are with us only in winter, that status has begun to change, as they adapt to our gardens and lawns. Nesting robins now are occasionally found south to Corpus Christi and Kingsville.

Robins are very special for several reasons, the least of which is their bright color and friendly manner. They are among the first birds our children come to recognize. And their wonderful songs provide us with ringing carols most of the year. When nesting, their songs are the first we hear in the mornings and the last we hear at dusk: "cheer-up, cheerily, cheer-up, cheer-up, cheerily." With a song like that, is it any wonder that "When the Red, Red Robin Comes Bob, Bob, Bobbin' Along" is our standard for happiness songs?

Most of us picture robins hunting earthworms in the yard, running here and there, then standing erect, cocking their heads, and when spying an earthworm, gradually pulling it out to haul it off to a waiting family. The old question of whether they find worms by hearing or by sight has conclusively been resolved. Cocking the head gives them a better view. Earthworms are favorite food during the nesting season but are seldom consumed during the remainder of the year.

Unlike most songbirds, once the young are fledged and accompanying the males, and after the female has completed the second brood, they all roost together, sometimes in congregations of several thousand. Their diet then changes suddenly from earthworms to berries, which they will eat until the following spring when they begin to feed their next brood. Although cherries are their favorite fruit—they will "belch out" the undigested seeds—bayberry, blackberry, greenbriar, hackberry, hawthorn, honeysuckle, juniper, mulberry, poison ivy, pokeberry, pyracantha, sassafras, sumac, and viburnum berries are all utilized. And many of us have seen robins succumb to the intoxicating effects of certain berries. They may flop, flutter, and stagger, sometimes even pass out, but by the next day they have recovered sufficiently to fly safely away, birdy hangover and all.

Another habit of robins is that of bathing. They seem to thoroughly enjoy this activity and will often patiently wait their turn to splash and chirp in what seems like pure contentment. I am convinced that our American Robin will bathe twice daily, in the morning and afternoon, whenever the opportunity allows. They will take advantage of almost every conceivable water source, from birdbaths to rain puddles, or even dew standing on leaves.

Our American Robin may be our true national bird, in fact if not by law.

It actively seeks our company, forages in plain view on our lawns, and nests in our shade trees. It is our most visible and cherished songbird.

Two Kinds of Meadowlarks Are Winter Residents

JANUARY 12, 1997

On clear, pleasant days in winter, one can hear the songs of both the resident Eastern Meadowlarks and the visiting Western Meadowlarks in our fields and pastures. Eastern Meadowlarks sing a plaintive series of two to eight whistle notes, often slurred and usually descending in pitch. Its song has been loosely translated as "spring is here" or "spring-o'-the-year."

Western Meadowlarks, however, sing a richer, lower-pitched song with two distinct parts: one to six whistle notes, followed by more complex, liquid, consonant-like notes. When growing up in the intermountain West, where Western Meadowlarks are common in summer, I was told that they sang "Salt Lake City is a prit-ty lit-tle city" or "Oh, yes, I am a pretty-little-bird." A great description of their beautiful song.

Wintering meadowlarks seldom sing their full complement of notes in South Texas, especially in early winter before the days begin to get a little longer and they start to think about things other than food and shelter. And so it sometimes is difficult to separate the two species audibly. They also are difficult to tell apart visually. But anyone with a good ear for sounds can separate the two species by the call notes that they utter on a regular basis. Our Eastern Meadowlarks possess harsh, rasping "dzert" or longer, buzzy "zeree" calls. The visiting Western Meadowlark calls are lower in pitch, sharp "chupp" or "chuck" notes.

Meadowlarks rank among our most common field birds, and are sometimes locally known as "Field Larks" or "Yellow Grassbirds." But in spite of their abundance, biologists may have incorrectly taken them for granted over the years.

Recent studies suggest that Texas may have three instead of two meadowlark species. The range of Eastern Meadowlarks extends from eastern North America west to the Canadian, Red, and Brazos Rivers (and south along the Gulf Coast), but it stops where the floodplain becomes narrow and the rivers are confined to rugged canyons.

Farther west, after a gap of about 80 miles, another meadowlark (currently regarded as the Lilian's race of the Eastern Meadowlark) breeds in the dry desert grasslands of West Texas, New Mexico, and Arizona, and south into Mexico. Western Meadowlarks also breed throughout this range (but they do not interbreed with the Eastern Lilian's form) and west and north into the western plains and grassy basins. Evidence from distribution, song, size, shape, and plumage suggests that the Lilian's Meadowlark is a full species.

Even this familiar and well-loved bird still has secrets that we are just beginning to examine. How many more are we yet to discover?

A Perspective on Winter

JANUARY 14, 1996

Winter seemed to have skipped South Texas in 1995, with temperatures reaching the freezing point or below only in a few scattered locations. But in 1996, winter's frigid conditions were felt on several occasions. While many folks appreciate the colder conditions, others would just as soon have warm, sunny days year-round. There is something to be said for both sides.

South Texas is situated in the southern latitudes, where our native plants and animals are well adapted to warm conditions with only occasional cold spells. Many of the exotic plants we decorate our yards with are from the south and can't stand freezing conditions. Conversely, northern plant species often require freezing conditions to germinate and produce flowers and seed.

While our native animals are adapted to the changing wintertime conditions, many of our exotic critters must be pampered if they are to survive a fierce norther. And our human neighbors' appreciation of winter largely depends upon where they were raised.

Officially, wintertime starts when the sun is at its southernmost position and the duration of daylight is at a minimum on December 21 or 22, the winter solstice. Thereafter, daylight and the angle of insulation gradually increase until once again there are twelve hours of daylight at the spring equinox on March 21. Winter generally includes an eighty-nine-day period from December 21 to March 21.

The South Texas winter includes the months of December, January, February, and, sometimes, March. The term *December* comes from the Roman's *tenth month* that was later appropriated as the twelfth month in the Gregorian calendar. *January* comes from the first Roman month *Januarius*, after Janus, the god of doors and gates. It refers to the entrance or way in—the start of the new year. *February* was taken from the Latin word *februare*, meaning "to make pure." The Old English name for February was *Sprout-Kale*, because in England cabbage begins to sprout in mid-January. It was later changed to *Solmonath* (meaning sun month) because it is the time when the sun rises higher in the sky and begins to drive away the chill of winter with its glowing rays.

March is Latin for *Mars*, the god of war. It refers to its rough and boisterous weather. Another name for it was *Lenctenmonath*, the lengthening month, because it is during March that the days rapidly become longer.

Most years, South Texas winters are mild, and we need a heavy coat only on the rare occasion when a cold front reaches into our area. By mid-February, spring usually is beginning to show, and then it is only a few days before our first Purple Martins arrive.

Raccoons—They Are Intelligent, Neat, and Dainty

JANUARY 15, 1995

Few mammals are as well known and admired, but also as despised, as the raccoon. It may be cute and fun to watch one minute and be cunning and divisive the next. In a very short time it can wipe out bird feeders or anything else edible left unattended.

Although raccoons prefer wetlands, they possess the capability to live practically anywhere in South Texas. They are omnivorous, able to eat almost anything alive or dead. Fish, frogs, snakes, snails, small mammals, and birds and their eggs are all susceptible. Even vegetable matter is regularly consumed, including mesquite beans, grapes, acorns, persimmons, cactus fruits, and all the berries they can find. In summer, they may even eat adult and larval wasps and their stored foods.

Raccoons are easily recognized by their robust, small-dog-sized appearance, black mask, and ringed tail. An extremely large 'coon may weigh fifty pounds. They are known to scientists as *Procyon lotor*, Latin for "dog that washes its food." This is because of its basic appearance and its eccentric habit of washing its food whenever possible. It may even carry food for a considerable distance to wash it before eating it.

Raccoons are intelligent, neat, and dainty.

Raccoons possess amazing dexterity. They are able to use their "hands" almost as well as humans do. Their fingers are long and extremely sensitive. They are used not only to grasp food but also to grasp branches when climbing trees, hunt for crayfish, open mussels, strip husks from ears of corn, and pick fruit and nuts.

There are few more amazing sights than an old raccoon crouched by the edge of a pool, looking elsewhere as if totally uninterested, while its busy fingers are exploring every nook and cranny under the bank for some frog that thought itself safe in the underwater retreat.

Raccoons normally den in hollow trees or logs but sometimes utilize cavities in banks or cliffs or even in old deserted barns and other structures. Daylight hours are usually spent sleeping because they are naturally nocturnal in their habits.

Their breeding season begins in February. A single litter of 1 to 7 (the average is 3 or 4) tiny youngsters are born in April or May. Females handle all of the family chores, from tending the young to teaching them the way of life after leaving the nest. The male refuses to assume any responsibility.

William Davis, in *The Mammals of Texas*, tells about a female that reared her three newborn in a "nail keg that had been fashioned as a nest site for wood ducks and wired 16 feet up in a tree standing in water 20 feet from shore." A racoon has also been found using a crow's nest as a daytime roost. In

Colorado, a mother and her "naked and blind young occupied a large magpie nest" in a scrub oak tree.

Raccoons are polygamous or promiscuous in their relations. Females reach sexual maturity in nine to ten months, but the males become sexually active only when about two years of age. After a brief midwinter courtship, he returns to a solitary bachelor's life.

Early Americans valued the raccoon for its meat, which has a good taste but is rather greasy. And its fur was famous for coonskin caps in frontier days. Although it is rarely eaten today, and coonskin caps are more a novelty than a practicality, raccoons remain one of our most abundant and fascinating wild creatures.

Red Admiral Is Our Most Common Wintertime Butterfly

JANUARY 18, 1998

Maybe it is because this winter has been so mild, at least so far, that I have seen a number of butterflies just about all winter long. Little Yellows, Cloudless Sulphurs, and Red Admirals are the ones seen most often. And American Ladies, Common Buckeyes, Pipevine Swallowtails, Texan Crescents, Gray Hairstreaks, and Fiery Skippers have been seen on occasions as well.

Of all these, however, none represent our winter butterflies as well as the brightly colored Red Admirals. Not only have I found these lovely butterflies in my own yard, but I have also found them in numerous other places throughout the winter, including the three areas in which I participated in Christmas Bird Counts: Coletoville–Upper Mission Valley Road area of Victoria County, Mad Island Marsh in Matagorda County, and Aransas National Wildlife Refuge in Aransas County.

Although some of those Red Admirals undoubtedly had been winter lingerers—individuals that pupated last fall and were able to find adequate shelter during the few days when the South Texas nightly temperatures dropped into the low 30s—others are more recent products. The fresh individuals are shiny and bright, truly spectacular in appearance. They are a velvet black with a red-orange bar across both wings and a broad red-orange trailing band on their hindwings. Their wingtips also contain contrasting white patches. They are moderate-sized butterflies (1 ¾ to 2 ½ inches), somewhat smaller than Monarchs but considerably larger than Little Yellows. Red Admirals can hardly be confused with any other Texas butterfly.

In summer, Red Admirals are widespread all across North America and into Central America; they even reach Alaska. And they possess one of my favorite scientific names—*Vanessa atalanta*—a name that somehow hits a romantic note in my psyche.

Another characteristic that gives Red Admirals a rather special appeal is their territorial behavior. Individuals actually stake out territories, where you can usually locate the same individual day after day, that they will protect from other butterflies. These aggressive creatures will usually dart out to challenge other animals, such as birds, mammals, and even humans that might be passing by. Sometimes they will chase other butterflies a considerable distance before returning to their territory. And they also may alight on one's arm or face to take salts from perspiration.

Red Admirals feed often at various flowers, although they seem to go for long periods of time just sunning themselves on the side of a building or on a tree trunk. During these periods, it may be possible to approach them to within a few inches if you approach very slowly without any sudden movements. Currently, they are nectaring on the whitish, bell-like flower of Viburnums, which are already blooming in my yard. And they also are feeding on an overly ripe banana that I have placed in an open feeding tray swung on a line from a Live Oak branch. One day, three individuals were perched on the gooey, brown banana, each with its proboscis deep inside the rotting fruit.

Flowering Viburnum, a nonnative plant that is available at most nurseries, is one of the very best plants for attracting butterflies during January. My native Agarito shrubs, which represent one of the earliest spring bloomers, have not even begun to bud, but my Viburnum flowers seem to attract most of the butterflies that are currently active. It is a real showplace!

An Eastern Screech-Owl Is Most Welcome

JANUARY 19, 1997

Finally, after I've lived seven years at my Mission Oaks home, an Eastern Screech-Owl has taken up residence. It is most welcome!

I put up a nesting box a couple of years ago, in hopes of encouraging some passerby to stay a while. And every now and then, either at dusk or dawn, I have whistled Eastern Screech-Owl songs, trying to convince one that he would be among friends. Until this week, my solicitations have been ignored. But now at least one screech-owl is singing in my backyard each evening, and on one occasion I actually observed it flying by at dusk.

Screech-owls are the smallest of our local owls, about 8½ inches tall, with an 18- to a 24-inch wingspan and a pair of feather-tufts that look like ears. Their plumage comes in a gray or reddish phase, and their breast is streaked with blackish lines. Their face is edged with black, and their eyes include a dark pupil and yellow iris.

They are most often detected by their usually soft quavering whistle calls. Since they normally are active only after dark, they rarely are seen during the daylight hours. Their songs may descend in pitch, or they can be a long, single trill all on one pitch. The trill song can accelerate in tempo. And they also possess a snappy, barklike whistle, used as an alarm call. Males and females sometimes sing a trill duet during the mating season; the songs of males are lower pitched.

Screech-owls nest in cavities, such as old woodpecker holes, cracks and crevices in trees, and even in nest boxes. Territorial behavior, including lots of singing, begins in midwinter and continues until their youngsters, usually four, are fledged in April or May. Their territories are selected based on available nesting sites and food. My yard, where I feed birds year-round, where the seed undoubtedly attracts mice, and where there are plenty of tall trees and shelter, contains all the necessary requirements.

Mice and other small rodents are only part of their diet, as screech-owls will eat almost any kind of small mammal, bird, lizard, snake, amphibian, and even fish, crayfish, spiders, insects, and earthworms. They are silent but effective predators. They hunt from an overhead perch, like a post or tree branch, swooping down on prey spotted from the perch. Although only about the size of a robin, they have extremely sharp talons and can easily pierce the skin of anyone handling a bird. It is best to leave them alone and to admire them for their marvelous songs and their place in our web of life.

Rufous Hummingbirds Are Braving Cold Weather

JANUARY 21, 1996

Even after freezing winter days, a few hummingbirds have remained at various sites in South Texas.

Although the vast majority of the hummingbirds that visited our area in the fall continued south to wintering grounds in Mexico or Central America, a few individuals remained behind. The common Ruby-throated Hummingbirds that summered here have all gone south, some as far as Costa Rica. And the larger Buff-bellied Hummingbirds, which also occur in South Texas during the summer months, have also gone south. Only a few of the more hardy species have remained behind.

Two immature male Rufous Hummingbirds have fed at hummingbird feeders in my yard since November. Each is gradually acquiring adult plumage, an orange-red gorget and orange-rufous back and underparts. And they now possess the typical high-pitched wing trill in flight that was missing when they first arrived.

Rufous Hummingbirds are hardier than most other hummers because they have adapted to more northern weather conditions. These colorful birds nest only in the far northwest, from northern Oregon and Washington east into Montana and north to Alaska. Most spend their winters in the mountains of Mexico, but a few wander northeastward and may remain at scattered localities all winter. Overwintering birds feed on available nectar and insects and are able to survive freezing conditions due to two amazing adaptations, torpidity and food storage.

Many hummingbirds actually are able to enter a deathlike torpor overnight, where their body temperature and metabolic rate are lowered and their breathing is reduced. Several other kinds of birds, such as swifts, nightjars, and our common Inca Dove, also possess this ability. Second, a few hummingbirds possess a well-developed crop in the esophagus where they are able to store insects, nectar, and/or sugar solution for later digestion.

The question obviously arises about the logic of maintaining hummingbird feeders year-round. While many folks take in their feeders after the fall migration, before winter weather arrives, to entice the remaining birds to continue south, a few hummers will remain wherever they can find an adequate food supply. So maintaining hummingbird feeders all winter is a good idea so long as the feeders are maintained throughout the winter. The remaining birds depend upon that daily supplement, and we should not disappoint them.

My Rufous Hummingbirds appear daily at my feeders almost at first light. I am enjoying their antics as well as being able to watch their change of plumage from the subtle green and whitish phase to truly orange-red and rufous.

Yellow-bellied Sapsuckers Common the Winter of 1998–1999

JANUARY 24,1999

It seems to me that Yellow-bellied Sapsuckers are more numerous this winter than usual. At least they are around my neighborhood at Mission Oaks. Three individuals seem to have made my yard the center of their activities. Almost anytime of day I can find one or two of these woodpeckers maintaining sap wells in the various Cedar Elm, Sugarberry, and Live Oak trees.

Yellow-bellied Sapsucker, a name that may sound like it was invented by a cartoonist, is most appropriate. Although members of the woodpecker family, sapsuckers differ from typical woodpeckers in several ways. Most importantly, their diet consists primarily of sap that they obtain from various trees and shrubs. Although they also feed on insects, which are most important for their developing nestlings, and fruit and berries, sap is essential year-round. In fact, sapsucker tongues have evolved with featherlike projections, enhancing their ability to lap the gooey sap. Woodpecker tongues possess barblike projections that allow them to probe into holes drilled into trees and shrubs and retrieve insects.

Sapsuckers, of which the Yellow-bellied species is but one of four found in North America, are exceedingly important members of various North American bird communities. They are often referred to as *keystone* species because of their importance in the wildlife community. In fact, a number of other species take advantage of the sapsucker's habit of maintaining sap wells. Several warblers, kinglets, and wrens actually utilize the flowing wells by licking sap, while warblers and other birds, such as flycatchers and hummingbirds, also take insects that are attracted to the sap. In a sense, the sapsucker sap wells are often the center of an active community.

Our wintering Yellow-bellied Sapsuckers can readily be distinguished from other woodpeckers by their medium size, red forehead and throat, black-and-white face, black chest, and yellow belly. The slightly smaller Ladder-backed Woodpecker, a year-round resident, lacks the red throat and black chest, although males possess a red cap; they possess a black-and-white streaked back, ladderlike markings. The larger Red-bellied Woodpecker is fairly common in the northern half of the South Texas region, while the similar Golden-fronted Woodpecker occurs only in the southern portion of our region; it is rare in Victoria County. Both of these larger woodpeckers have a barred back and grayish underparts. In addition, the Northern Flicker is also fairly common in winter; it is easily identified by its yellow flight feathers and spotted breast. The much larger Pileated Woodpecker is a bird primarily limited to riparian areas, and the Red-headed Woodpecker is only rarely found in South Texas.

Yellow-bellied Sapsuckers only winter in South Texas, usually arriving in early October and departing in April. Evidence of their activities is easily found by the lines of small test holes on the trunks of various trees and shrubs. They maintain only the most productive ones. Their nesting grounds lie far to the north of Texas, across the northern tier of states and northward into Alaska and east to Newfoundland. They overwinter throughout the southeastern states from most of Texas eastward to the southern Atlantic states, and a few occasionally reach the Caribbean Islands.

Their presence in our neighborhoods is worth noting. They are a fascinating species and truly an important member of the wildlife community.

Plan Your Hummingbird/Butterfly Gardens Early

JANUARY 26, 1997

Start looking ahead to what shrubs and flowers you will plant to attract hummingbirds and butterflies come spring. And learn from the recent freezes as to what plants are most likely to survive South Texas winters. I must admit that I haven't done a real good job in past years, as many of my preferred species didn't make it through the winter. But who was it that said the key to success is to learn from one's mistakes? I should be really smart!

I also have an advantage because I have been paying a lot of attention to the almost daily discourse on the Internet about what hummingbird plants did best the winter of 1996–1997 in Louisiana, where their winters are pretty much like those in South Texas. Anyway, I will share what I have learned of late about the hardiness of various plant species.

After 20-degree temperatures, Flowering Maple or Orange Abutilon (*Abutilon pictum*), Shrimp Plant (*Justica brandegeana*), Cardinal's Guard (*Pachystachys coccinea*), Carolina Jesmine (*Gelsemium sempervirens*), Winter Honeysuckle (*Lonicera fragrantissima*), Common Camellia (*Camellia japonica*), and Brazil Sage (*Salvia guaranitica*) showed no damage; Coral Honeysuckle (*L. sempervirens*) and Smuggler's Vervain (*Stachytarpheta sp.*) showed only moderate damage.

The following species, however, all considered good for hummers, were "losers": Mexican Bush Sage (*Salvia leucantha*), Pineapple Sage (*S. elegans*), Tropical Sage (*S. coccinea*), Texas Mallow (*Malvaviscus drummondi*), Turk's Cap (*M. arboreus*), Firebush (*Hamelia patens*), Powderpuff (*Calliandra haematocephala.*), Cigar Plant (*Cuphea ignea*), Firespike or Red Justica (*Odontonema strictum*), Firecracker Vine (*Manettia cordifolia*), Pagoda Plant (*Clerodendrum speciosissimum*), Japanese Honeysuckle (*Lonicera japonica*), and Four O'clock (*Mirabilis jalapa*). A few of these, such as Turk's Cap, Firebush, and Cigar Plant, are certain to come back, even after extremely hard freezes.

A number of good butterfly/hummingbird plants do well over winter in planters; these can be moved indoors or outdoors with the weather. I have successfully grown Pentas (*Penta sp.*) and Fountain Flower or Coral Plant (*Russelia equisetiformis.*) this way; it is amazing how the Pentas readily attract passing butterflies and the Fountain Flowers attract hummingbirds.

The above list of plants is only a start in helping to select those that are most attractive to hummingbirds and butterflies. For instance, a number of herbs (available in spring) are great butterfly attractants: dill, fennel, and parsley. Marigolds, periwinkles, and zinnias are popular as well. And Cardinal Flower, Cherry Sage, columbines, Crossvine, Desert Willow, Red Yucca, Trumpet Vine, and Tree Tobacco attract hummers wherever they occur.

Its time to think spring. Get with it!

Mexico and South Texas Share Strong Ties in Nature

JANUARY 28, 1996

On a trip to Mexico, I participated in the Gómez Farías, Tamaulipas, Christmas Bird Count on New Year's Eve of 1995. My visit reminded me of the close biological ties between northern Mexico and South Texas.

I spent several days in the Rancho del Cielo cloud forest, considered the northernmost outpost of that habitat. I also visited several areas in the lowlands. And I drove out to the beach near Matamoros and wandered along two great Mexican rivers, the Río Corona and the Río Sabinas, about 150 miles south of the border; both originate in the Sierra Madre Oriental. The Río Corona flows into the Soto La Marina and then into the Laguna Madre at La Pesca. The Río Sabinas flows into the Río Tamesi that enters the Gulf of Mexico just below Tampico.

Matamoros's Playa Baghdad, at the end of Highway 2, looks all the world like a Texas beach. After all, it is protected by the same system of barrier islands that occurs along most of the Texas Gulf Coast. The same plant life occurs throughout, and Texas shares the same coastal birdlife.

The Rio Grande provides not only a line of greenery, where many of our common wildlife find shelter and food, but also an important conduit for nutrients between the interior and the Laguna Madre. The outflow of the Rio Grande provides great feeding grounds for multitudes of marine life from both countries.

Further south, I found thousands of geese—primarily White-fronts and Snows, along with fewer numbers of Canada and Ross's Geese—feeding in the agricultural fields, just like they do in Texas's multicounty area. These waterfowl will move back and forth with the weather conditions and available food supply.

The Río Corona and Río Sabinas, like the Rio Grande, are significant conduits to the Laguna Madre. Huge Montezuma Bald Cypress trees still occur along these rivers, although much of the adjacent areas have been converted to agricultural uses. These riverways are even reminiscent of the Guadalupe riverway, starting with the dominating Cypress trees and the abundance of birdlife. However, a number of tropical species, such as trogons and motmots, reach their northern limits at the Río Corona; temperate species, such as the American Crow and Carolina Chickadee, reach their southern limits at the San Antonio River.

What is most interesting about the river corridors of the Guadalupe, San Antonio, Rio Grande, Río Corona, and Río Sabinas, north to south, is the changing birdlife. South Texas now claims several bird species that, until the last two decades, only occurred below the border. Now some of the tropical species—Muscovy Duck, Hooked-billed Kite, Ringed Kingfisher, Brown Jay, Tamaulipas Crow, and Clay-colored Robin—are resident in the lower Rio Grande Valley. Earlier tropical arrivals included the White-tipped Dove, Northern Beardless Tyrannulet, and Tropical Parula.

The reason for their appearance in Texas is unclear: global warming, habitat loss south of the border, or simply the birds' ability to expand their ranges. One can't help but wonder how long it will be before some of these birds will occur regularly along the San Antonio and Guadalupe rivers.

Male Wood Ducks Are Our Most Beautiful Waterbird

JANUARY 29, 1995

Anyone who has not seen a male Wood Duck has missed one of nature's most spectacular creations.

Males possess an almost surreal appearance in their rather gaudy attire of blue, burgundy, white, black, and buff. The long, iridescent green-to-purple crest is highlighted by a white line that runs from the back of its red bill to the tip of the crest, extending halfway down its burgundy-colored neck. A second white line starts at the base of the crest and runs from behind its blood-red eye to the tip of the crest. Additional white lines extend from its snow-white throat onto the cheeks and halfway around the neck. The Wood Duck hen, on the other hand, is a rather drab, mottled brown bird with a large, elliptical, white eye ring.

Wood Ducks are surprisingly common at quiet ponds and along protected riverways in South Texas. But they are not easy to see well unless one approaches the habitat in a very slow and cautious manner. A sudden and noisy approach will immediately frighten them off, and your view will be limited to retreating birds calling their characteristic "whoo-eek" flight calls.

It is possible to get reasonably close if the ducks do not think they are being observed; keep your glances away until you get close, and then keep shrubs or trees between you and your quarry. They often will freeze in place, and you can get a wonderful view with binoculars. And on rare occasions they can be found out of the water, where they are searching for insects and seeds, including acorns. In ponds, they feed primarily on duckweed.

Wood Ducks are early nesters, utilizing natural cavities in trees and other structures, as well as boxes placed in suitable sites near out-of-the-way wetlands. The hen leads the search for a proper nest, with the male tagging along. The hunt may take several days.

Normal clutch size is about 12, although as many as 50 eggs have been found in a single nest, with many of the eggs contributed by other hens. Frank Bellrose, author of *Ducks, Geese & Swans* of North America, claims that a "nest with over 25 eggs is generally conceded to be a dump nest." As many as five Wood Duck hens may deposit eggs in a single nest, although only one hen does the incubation.

The drake remains involved only until all the eggs are hatched. He then leaves to join bachelor flocks in more secluded places. It is left to the hen to entice the chicks, which are equipped with claws for climbing, out of the nest and down to the ground, and then to lead the family, often through difficult terrain, to the greater security of ponds. The protection and training of the youngsters also are left totally to the hen, who often will receive assistance from other hens. Chick-sitting, when the principal hen goes off to feed elsewhere, is also a common practice.

Our Wood Duck may be the most exquisite of American ducks, displaying all the colors of the rainbow in its plumage and a touch of poetry in its scientific name, *Aix sponsa*, a mixed Greek and Latin term meaning "waterfowl in wedding raiment."

Purple Martins arrive in South Texas in February.

February

It's Time to Get the House in Order for Purple Martins

FEBRUARY 2, 1997

Still winter and too early to get ready for Purple Martins? Think again. Some of the first males can arrive in South Texas the first week of February.

Although our first martins usually do not appear until mid- to late February or early March, it's time now to prepare for their arrival. That means cleaning the martin house, getting rid of the spiderwebs and insects that may have taken over since the rightful tenants vacated in midsummer of last year, and hoisting it up the pole so it's ready and waiting.

More often than not, martin house preparation also requires a fresh cover of white paint. The light color helps to reflect the hot Texas sun and also to highlight the entrance holes.

In case this is your first time at attracting martins, here are some easy rules to follow:

* Houses must contain apartments with at least a 6 × 6-inch floor space and an entrance hole 1¾ inch in diameter and 1 inch above the floor.
* Houses must be placed on poles 12 to 20 feet above the ground and should be 40 feet away from taller trees, poles, and other structures.
* Poles must be free of vines and shrubs that might allow access to the house by predators.
* Houses must be free of nesting materials and other debris that accumulated in the off-season.

Purple Martins often are rather finicky at the start but seem to put up with shorter poles and poorly maintained structures once the colony is established. Most birds are repeats, but the majority of the first-year birds (usually last year's youngsters) seek out new sites, usually in the general area of their natal homesite. This means that a new martin house, especially if it is in the proximity of an active martin house, is likely to be used early on. Distant houses are not as likely to be selected.

Another way to attract first-year martins is to play a tape of their dawn chorus. Playing Purple Martin songs at a new martin house will certainly attract their attention. And if they like their new digs, they will probably remain and nest. If not, give it time, and sooner or later you will attract martins that will begin a new colony.

An established Purple Martin colony is likely to return year after year so long as you maintain the house and environment. They will consume millions of flying insects during the short time they are with us. And they will also provide us with their marvelous songs from long before dawn to throughout the day and evening. But by mid- to late July they will leave our neighborhoods and begin their 5,000-mile southward migration to their wintering grounds in South America.

But rather than think about their departure, think first about their arrival. It is time now to prepare. Good luck!

Common Grackles Are Already among Our Oak Trees

FEBRUARY 5, 1995

The warmer-than-normal winter days have already begun to produce a crop of tiny Oak Moth caterpillars, which normally do not appear for another month or so.

How do I know? No, I haven't crawled up into one of the Live Oaks that are so abundant in South Texas, but I have seen flocks of Common Grackles in the oaks in my neighborhood. These all-black birds are all-knowing when it comes to finding food, and the fresh crop of insects in the oaks is all it takes for them to change their habits from foraging on the ground to gleaning insect larvae in the oaks.

We have two kinds of year-round resident grackles in South Texas. The first is the Great-tailed Grackle, identified by its extremely long tail. The glossy purple-black males possess yellowish eyes. The second is the smaller Common Grackle, with a glossy purplish head that contrasts with its brownish back and belly. Also, in late spring and summer, a third grackle—Boat-tailed Grackle—is fairly common along the coast.

Although the three species can look very much alike, since they look all black at a distance, the Common Grackle has a very different personality than its two longer-tailed cousins. Common Grackles are truly woodland birds, while the Great-tail prefers open pastures and wetlands.

Although Common Grackles spend considerable time on the ground, as do Great-tailed Grackles, they rarely are found away from the protection of the woodlands. They walk about sedately, often nodding their held-high heads, and they may actually dig out insects and seeds with their long, tapered bills. Their diet includes almost anything they find, from insects and spiders, to snails and earthworms, to an occasional fish, crayfish, frog, snake, or mouse. Weed seeds, acorns, and variety of grains are also favorite foods. Biologists have reported that 66% of a grackle's diet consists of beetles, including Japanese beetles and several other species that eat our crops.

All the while they are searching for food, whether in the oaks or on the ground, grackles produce a wild assortment of sounds, ranging from low chucking noises to crazy combinations of squawks and squeaks. And when disturbed, by either a neighbor or a predator, they may open their wings, spread their tail, and puff out their plumage to make themselves appear larger and fiercer than they really are.

Like blackbirds everywhere, Common Grackles, which sometimes number in the thousands, fly to favorite roosts for the night. They may even roost with other blackbird species, such as other grackles, Red-winged Blackbirds, and cowbirds. But at dawn, they all go their own way to search for food in their own peculiar ways.

Birds Love to Bathe in almost Any Weather

FEBRUARY 7, 1999

Bathing birds seem to create a bathing frenzy. It seems to me that if one bird begins to bathe, suddenly all those within sight or hearing distance also get the urge to bathe. I recently watched as one cardinal began to bathe in a birdbath in my yard. Two Chipping Sparrows that were searching for seeds on the ground nearby immediately joined the cardinal. And within the next two or three minutes they were joined by more cardinals and Chipping Sparrows, two American Goldfinches, a White-throated Sparrow, and a pair of Inca Doves.

On numerous other occasions, I have noticed this sudden urge for birds to bathe, no matter the time of year or the temperature. I have even seen robins bathing in water running off snowbanks in Utah. There seems to be some magical moment when all the birds suddenly decide it's bath time.

Bathing is a normal behavior in all birds; almost all species bathe. And if water is not available, such as in very arid areas, many species will take dust baths. Both kinds of bathing are undertaken for a number of reasons, including, perhaps, just for the joy of it. But bathing serves two functions for birds. First, it is important for feather maintenance, by cleaning them so they do not get so oily and dirty as to hinder flight. The flight-feathers on the wings are most important, since birds depend upon their wings for movement of all kinds, including escaping from predators and for migration. Second, and also extremely important, bathing helps clean the body and feathers of parasites that might otherwise become so numerous as to create health hazards.

How often do birds bathe? It depends a lot on the weather. On hot summer days they bathe more often, sometimes as much as four or five times a day, than on cooler winter days. And what is equally important is the shaking and preening that takes place after each bath. A bird will stand still and shake its feathers, usually with wings drooping, for several minutes after each bath. It also will run each feather through its bill, actually nibbling along the feather to take off each bit of debris and to interlock the feather-barbs again.

Bathing techniques, however, can vary considerably. Typically, a bird will enter the shallow water, stand a few moments, and then begin dipping and throwing water over its back with its wings. This may continue for three or four minutes. Robins, sparrows, cardinals, warblers, and doves bathe in this fashion. Vireos, such as our common resident White-eyed Vireo, will rarely enter the water per se. Vireos tend to dive down from a few inches above the water and dip the water over themselves, little more than a spit bath.

Terns and lots of other aerial feeders dive into the water from a few feet high and immediately take off, shaking the water off. Gulls and rails bathe while wading.

After a bath a bird often will perch on an open snag or branch and dry off by spreading its wings. We see this spread-wing behavior in vultures during the early morning hours, but that behavior is intended to warm up the bird after a cool night. Wing-spreading after a bath is a drying procedure. Anhingas and cormorants are well known for this after fishing, but unlike most waterbirds, anhingas and cormorants require this procedure because they lack the oil glands that most waterbirds use to oil their feathers.

The best bird attractant in any yard is a birdbath. Only a certain percentage of our resident and visiting birds will come to a seed feeder, but all birds bathe. Splashing water, either from a dripper or from other bathing birds, is a great attractant. A birdbath is a must in any yard.

Which Early Birds Will Be the First to Arrive?

FEBRUARY 8, 1998

Already a few of the early migrants are beginning to appear in South Texas. Purple Martins have been spotted at a number of places, and Ruby-throated Hummingbirds have returned to various feeders. I have at least two subadult male Ruby-throats that have been present for several weeks; maybe they never went south this year. And for the first time at Mission Oaks, where I live near Mission Valley, Buff-bellied Hummingbirds have stayed all year. Everyone agrees it has been an extremely mild and out-of-the-ordinary winter. What other early migrants can be expected?

Shorebirds are classic early migrants. Although many species overwinter in our area wetlands and along the coast, others, including American Golden-Plovers and Upland and Pectoral Sandpipers, are early migrants only. Start watching for these shorebirds anytime now; these are grassland birds that frequent our short-grass fields and pastures in passing during their journey north. Other migrant shorebirds, such as Hudsonian Godwit, White-rumped, Baird's, and Buff-breasted Sandpipers, and Wilson's Phalarope, move through South Texas a little later.

What will be our earliest songbirds this year? Typical early migrants include Great Crested and Scissor-tailed Flycatchers; Western Kingbird; Rough-winged, Bank, and Barn Swallows; Yellow-throated Vireo; Nashville, Yellow-throated, Black-and-white, and Wilson's Warblers; Louisiana Waterthrush; Yellow-breasted Chat; Indigo Bunting; Yellow-headed Blackbird; and Baltimore and Bullock's Orioles. If the past is any indication, instead of showing up in mid- to late February, these neotropical migrants (species that winter in the tropics and nest in North America from Texas north to Alaska) may be found earlier in the month.

Of course, the big push of migrants usually doesn't begin until March or early April, when many of South Texas's year-round residents are already nesting. Although some of these early migrants, such as the Scissor-tailed Flycatcher, remain and nest, other summertime residents usually arrive in this later group. One of these, and one that may signify true spring to many bird lovers, is the Cliff Swallow. It seems that spring has arrived only once they are back. These gregarious birds utilize various man-made structures, such as overpasses and concrete bridges, on which to construct their mud-pellet nests. Some colonies may number in excess of two hundred birds.

My favorite time in spring is mid- to late April, when the greatest mass of northbound migrants—as many as one hundred or more species in a single day—are passing through South Texas. Although the greatest numbers of birds can usually be found along the coast, especially on the barrier islands, inland sites can also be filled with varied and often colorful species. But weather conditions are largely responsible for the numbers.

Stormy conditions over the Gulf tend to exhaust the Trans-Gulf migrants (birds that cross the Gulf of Mexico instead of following a land route). Bad weather drives them westward to the Coastal Bend. During good weather, especially when there is a breeze out of the south, the majority of the Trans-Gulf migrants will pass over South Texas altogether or remain to the east, landing along the northern Gulf Coast or even further inland. As strange as it may seem, birders hope for stormy conditions in spring over the Gulf.

But whether you are an active birder, one that braves the stormy weather to find as many birds as possible, or simply a backyard birder who enjoys watching the birds that frequent your feeders, spring is a time to anticipate new birds and increased activity. And some of the songbirds that you may see will undoubtedly provide added pleasure in being one with nature.

Texas Bluebonnet, Texas state flower.

Spring Ushers in the Return of Texas Bluebonnets

FEBRUARY 11, 1996

Already some of South Texas's earliest spring flowers are starting to bloom, and none are more welcome than Texas's own bluebonnets.

Not only is the Texas Bluebonnet our official state flower, but it unquestionably is among the most widely recognized flowers in the United States. It has become synonymous with clean and colorful roadsides, primarily due to the extensive public relations campaign undertaken by the state of Texas, but also due to the broadscale program of roadside beautification that was started and supported by Lady Bird Johnson.

Almost everyone recognizes the Texas Bluebonnet as an upright plant with deep purple-blue flowers tipped with silvery white. Its green leaves are divided into five leaflets. Bluebonnets readily grow along our highways, usually on slightly drier sites, and oftentimes huge fields are dominated by their brilliant blooms.

The Texas Bluebonnet is a member of the widespread legume or pea family, as are acacias, mesquites, and locoweeds. The abundant "Texas" Bluebonnet occurs only in Texas, within a north-south swath across the eastern half of the state. Four other bluebonnets, all of the genus *Lupinus*, occur in Texas, including the unique Big Bend Bluebonnet of the Big Bend area.

The genus of *Lupinus* includes approximately two hundred species that are scattered all across the temperate regions of the world in both hemispheres;

exceptions include Africa and Australia. And, although they all have similar flowers and leaves, they vary from diminutive forms in the western deserts to huge tree-lupines in the tropical mountains. On Sierra Colima in southwestern Mexico, those lupines can get ten feet tall. I once found several colonial-nesting Gray Silky-Flycatchers utilizing those lupines that were growing along the edge of the ash field below the smoking volcanic cone.

But none of the world's lupines are as bright and colorful as the Texas Bluebonnets that grace our highways and fields during spring. They usually last for several weeks, or until the surrounding grasses and other wildflowers grow higher and shade them out. Then we must wait for next spring's stupendous wildflower show.

Long-Legged Skeeters Are Actually Crane Flies

FEBRUARY 15, 1998

Even before the end of January this year, I discovered a few Crane Flies on the walls of the shed in my backyard. Another sure sign of an early spring!

I could find only about a dozen individuals, and they were in pairs, rather than in the huge numbers that occur at times. The great swarms that may occasionally be found in cool, damp places, such as in culverts or under concrete bridges, are, perhaps, more typical. And sometimes these swarms bob up and down, raising and lowering their bodies by bending their long legs in a really strange manner. This behavior is not well understood, although some entomologists suggest that it may be a way for the males, which dominate the swarms, to attract passing females. Others suggest that it is more related to safety in numbers. But whatever the reason, it is a little odd and also a little spooky.

Crane Flies are those huge, mosquito-like insects that, when flying about, are slow and clumsy, sometimes even bumbling into lights, doors, and even the mouths of predators. These daddy longlegs of the air have also been called the "drolls of the insect world" for their unassuming personality. They are not at all like most other members of the Diptera, or the true fly family, which includes such well-known creatures as the House Fly, mosquito, Fruit Fly, midge, and gnat. Most of these other flies are sun-loving creatures. There are about 90,000 Diptera species worldwide, including almost 17,000 in North America. Dragonflies, Mayflies, and Stoneflies are not true flies.

All of the true flies possess a single pair of wings and a pair of short, knobbed projections called *halteres*, located on their bodies just behind the wings, that serve as balancing organs. They are easy to see on Crane Flies, due to the insect's large size. Halteres act as a second pair of wings, like gyroscopes, vibrating rapidly in opposition to the insect's wing beat; when the wings move up, the halteres move down. If one of the halteres is removed, the insect can no longer fly; it sideslips and yaws out of control. In people, this sense of balance derives from structures in the inner ear. If something goes wrong with this mechanism, a person has difficulty in navigating and even standing up.

Crane Flies, unlike their mosquito cousins, have no sting or bite; they are totally oblivious to humans. One usually can get within a few inches for a close-up examination. They can be described as having a narrow abdomen, narrow wings, and absurdly long legs. Occasionally they can be found walking about on tree trunks, logs, or damp leaf-cluttered ground. Many of those found on the ground or on logs are males in search of a female. They may even sit beside a pupa until the female emerges and then mate, scarcely before she has freed herself of the pupal skin. The female Crane Fly deposits her eggs on the surface of rotting wood or pushes them into the soft pulp. The larvae, tiny greenish grubs, crawl about below the surface of the ground, feeding on roots and seedling plants, sometimes killing them. Although adult Crane Flies are most obvious and attract our greatest attention, the larvae, which are rarely evident, are biologically more important.

In spite of the relative unimportance of the adult Crane Flies, they are far more interesting at this stage than at the larval stage. Finding a swarm of these long-legged insects on some damp structures, or seeing several individuals flying about one's property on a warm spring day, seems to be a telling signal that the new season has truly arrived.

Early Signs of Spring Are All About

FEBRUARY 16, 1997

Already there is a scent of onions in our parks and similar grassy areas on warm days. Various citrus trees and forsythia shrubs have also begun to bloom.

A few Purple Martins are being seen along the coast, and these harbingers of spring will occur farther inland any day. The earliest blooms of Indian Paintbrush are appearing along the roadsides. And early Agarito shrubs are producing yellow flowers that attract a multitude of insects, especially bees, wasps, and butterflies.

The early spring butterflies include a few that have managed to overwinter in protected sites as well as those that have emerged from their winter chrysalis, the stage between larva and adult. During warm days in late January, I have found a few Gulf Fritillaries, Cloudless Sulphurs, Little Yellows, Snout Butterflies, Gray Hairstreaks, Texan Crescents, and Red Admirals.

In my yard, only Coral Honeysuckle and viburnum are blooming, following the several freezes that knocked back other flowering vines and shrubs. But these two plants, especially the white-flowering viburnum, are real magnets to early nectar-eaters. My hummingbirds, however, seem to prefer the feeders and the abundant insects that are numerous now among the Live Oaks.

Three kinds of hummingbirds have been utilizing my handouts since about Christmastime. The most dominating hummer is an adult Rufous Hummingbird, easily identified by its rufous rump, tail, and crissum. It chases all the other hummers away from whichever feeder it prefers for the day. A second Rufous Hummer, an immature male, is fast developing its adult features.

The largest of my yard hummers is an immature male Broad-tailed that has a broad tail with only a hint of rufous above the black-and-white tips and green center. And in the last few weeks, an immature male Ruby-throated Hummingbird has appeared. This bird has a distinct triangular red throat-patch and all green back and tail.

I am now starting to watch for our summertime hummer, the larger Buff-bellied Hummingbird. This tropical species usually appears in my yard during the first week of March, but with the strange weather we are having this year, I am monitoring my hummer herd for an early arrival. Buff-bellies are easier to identify than other hummers: they are larger than our other summertime hummer, the Ruby-throat, and possess a bright green throat and chest, buff belly, unspotted tail, and a red bill with a black tip. Truly an attractive hummer!

I guess my spring will not truly arrive until the Buff-bellies are back on territory.

Various Factors Contribute to Declining Cardinal Count

FEBRUARY 18, 1996

A lady in Ganado recently wrote me about "the diminishing numbers of cardinals in our area." She pointed out that in year's past up to fifteen had visited her feeders in winter, but now "we're lucky to get one or two." She wrote that some folks had blamed the decline on an increase in raccoons.

Raccoons do prey on cardinals and on any other birds (and their eggs) that they can capture, and raccoons are smart and opportunistic critters. There is no doubt that an increased population of raccoons can pose a threat to our songbird populations. However, keep in mind that these native animals, as well as numerous other potential predators, have survived together in the wild for as long as they have existed. And there still are plenty of "redbirds" throughout their very extensive dual range, the entire eastern half of North America.

The reason for a sudden decline in our Northern Cardinals can more likely be explained by one of several other factors. Here are some possibilities:

(1) Poisoning by chemicals used as insecticides or herbicides; many chemicals used regularly can have serious consequences for our wild birds and other wildlife.

(2) House cats, whether wild or maintained indoors, are a very serious threat to wild birds, especially ground-feeders like cardinals, robins, doves, and sparrows. It has been estimated that there are about 55 million house cats in the United States. Subtracting those that never go outside or are too old to hunt leaves about 44 million on the loose. If only one in ten of those house cats kills only one bird a day, this would amount to a daily toll of 4.4 million dead songbirds. Many house cats kill multiple birds a day.

(3) Diseases are also possible, but unlikely. Cardinals are amazingly rugged creatures that can withstand all sorts of natural threats.

Unkempt feeders can harbor bacteria and viruses that can affect birds that utilize the feeders. Feeders need to be cleaned and dried on a regular basis.

(4) All animal populations are cyclic in nature. Whenever a population increases to some magical size and the available water, food, or territory becomes too small, either that population divides, with some individuals moving elsewhere, the weaker individuals being excluded and eventually dying, or the population is affected by disease.

Only human beings, of all the animal kingdom, think they can beat the odds. But in nature, the population of every living organism has its limit. When any population (humans included) reaches its maximum size, like too many rats in a cage, there is only one consequence. Nature will always win in the end!

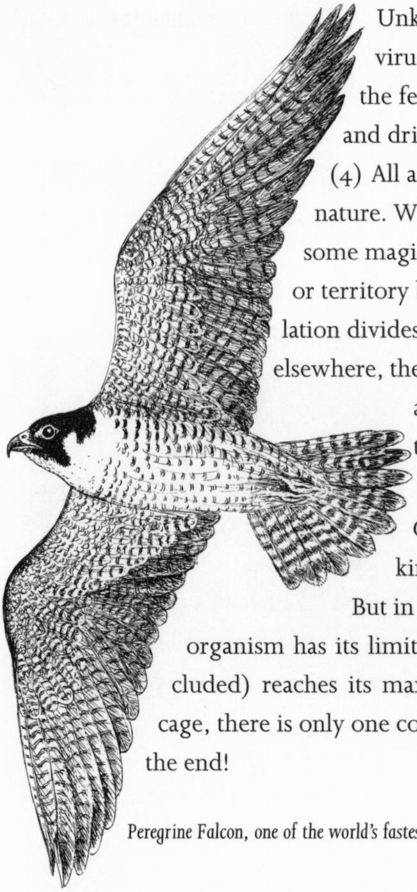

Peregrine Falcon, one of the world's fastest birds.

How Fast Do Birds Fly?

FEBRUARY 19, 1995

A Peregrine Falcon is generally acknowledged as our fastest bird. This large and powerful raptor has been clocked at over 140 mph and perhaps as much as 200 mph in a power-dive in pursuit of prey.

Peregrines reside through winter along the Texas coast, and occasionally they are seen thirty to fifty miles inland. Waterfowl and shorebirds are favorite prey species that, even in an accelerated escape mode, are in no way able to escape from the much faster peregrine. Ducks have been recorded to fly from 40 to 65 mph, while a Spotted Sandpiper, a common wintering shorebird, has been clocked at 25 mph.

Raptors are some of our fastest birds; they depend upon their flight speed to capture their prey. Pursuit predators like the Sharp-shinned and Cooper's Hawks may fly as slow as 16 to 20 mph but can quickly accelerate to 60 mph. American Kestrel flight speeds vary from 22 to 36 mph. Osprey speeds vary from 20 to 80 mph, and Golden and Bald Eagles from 28 to 44 mph. And the abundant Red-tailed Hawk commonly flies at 20 to 40 mph.

Hummingbirds seem to be extremely fast, but that is largely due to their relatively small size. Ruby-throated Hummingbirds, the common species that we see daily during the summer months, normally fly at 28 mph, although one was clocked at 50 mph with a tailwind. A hummingbird is unique for its ability to move its wings forward and backward in a figure eight, getting most of its driving power and lift from the downbeat. Like living helicopters, they are able to hover and fly forward, straight up and down, and backward. High-speed photography has been used to time their wing beats at about 70 times a second.

Wing beats also vary considerably, from the 70 per second of hummingbirds to as few as 2 beats per second for the American Crow. Chickadee wing beats average 27, Northern Mockingbirds 14, American Goldfinches 4.9, European Starlings 4.3, Eastern Bluebirds 3.1, Rock Doves (domestic pigeons) 3, Mourning Doves 2.45, and American Robins 2.3. And the Great Blue Heron, a huge wader that is common along area waterways and wetlands, barely beats its wings at a rate of one per second. Its normal flight speed has been clocked at 28 mph, but it has the ability to speed up to 45 mph. And Racing Pigeons (Rock Doves) have been officially timed at up to 82 mph.

But what about the flight speed of the average songbird that we see daily in our fields and gardens? Most fly at 15 to 25 mph, but they usually can accelerate to about twice that speed for short distances when pursued. The American Crow, among the most common birds in the Golden Crescent, has a flight speed of 30 to 45 mph; the common House Sparrow is one of the slowest at 16 to 19 mph; Blue Jays normally fly at 20 mph, Loggerhead Shrikes at 28, American Robins at 32, Horned Larks at 54, starlings at 55, and Barn Swallows at 60.

A bird's relative flight speed is most evident when it is fleeing from a faster bird of prey. Not long ago near Seadrift, I watched a Peregrine hit a shorebird from above. It was like a bolt of lightning from the blue. A flock of dowitchers, flying along at about 30 mph, probably never saw the Peregrine that suddenly descended upon them, hitting and instantly killing one of the flock. In an instant, the Peregrine had rolled, turned back, and took the falling dowitcher in midair. Amazing dexterity! It flew away to a quiet place to consume its kill.

Huisache Trees Are Blooming!

FEBRUARY 22, 1998

The widespread blooming of the little Huisache tree is one of our best indicators of spring. Its deep, rich yellow flowers—spherical heads about ⅔-inch across on 1- to 1½-inch stalks—are beautiful and aromatic spring heralds. They are especially noticeable because, rather than putting on a few scattered flowers that could easily be missed, hundreds of flowers usually bloom at once. An incredible show! And the bright yellow glow may last for several days or several weeks.

Huisache trees rarely occur alone but usually are found in numbers, all the better to brighten the landscape. They seem to prefer depressions, such as old stock tanks or resacas, places that hold water at least during parts of the year. They also appear on land that has been recently cleared, providing early vegetation that helps to limit erosion. These fast-growing trees can produce flowers in their second or third year. And if a great patch of bright yellow Huisache trees isn't enough to attract one's attention, their strong scent may also help.

This almost overpowering aroma has given the Huisache tree its common name of "sweet acacia." And because of the tree's aroma, it has become famous worldwide. In fact, it was highly prized in European gardens long before the first Texians reached their adopted country. As early as 1611, Cardinal Odoardo Farnese was cultivating Huisache plants for their aroma. It was Farnese who arranged for publication of his botanical treasures in 1625, in which the little American tree was described as *Acacia Farnesiana*, after the cardinal. More recently most botanists have begun to use the scientific name of *Acacia smallii*.

Later in the seventeenth century the tree was introduced into France, where the flowers were utilized as a base for Grasse perfumes. Huisache plantations were established for *cassie*, as it is known in France, and a local strain that produces two flower crops each year was developed. The odor is extracted from the flower oils and concocted into a pomade which goes into extracts of violet and aromatic vinegar to produce a very concentrated material known as *quintessence of cassie*. It is one of the most costly of all scents.

Huisache, pronounced "wee-sach" or "wee satch-eh," is a small tree or large shrub, rarely more than 25 feet tall, with a spreading rounded or flattened crown. However, the national champion Huisache, near Rio Grande Village in Big Bend National Park, measures 48 feet tall with a 60-foot crown spread and a 6.1-foot girth. Like all the other legumes (members of the pea family), Huisache trees sport sharp spines; they are paired, straight, 1–3 inches long, and appear at the base of each leaf. Also like most other peas or legumes, woody pods appear in summer or early fall.

The dark brown to black pods are 1–2 inches long and contain two rows of shiny, hard, gray seeds. Unlike mesquite pods, they are rarely utilized for food. However, according to Paul Cox and Patty Leslie in *Texas Trees: A Friendly Guide*, the "pods were formerly made into ink, the juice was used as a glue for mending pottery, and the bark for dying skins." What's more, "decoctions from the green fruit serve as an astringent and the roots were used as a treatment for tuberculosis. Wound dressings were made from the crushed leaves, and the flowers were used as an infusion for ingestion and as an ointment for curing headaches."

Today our lovely Huisache trees, rarely used at home for their assortment of benefits, but so well known elsewhere in the world, are among the greatest joys of spring. They are welcome harbingers of the new, fresh season.

Signs of Spring's Arrival Are All Around

FEBRUARY 25, 1996

Springlike days do not necessarily signify the arrival of spring. Our fickle weather can easily bring us more cold, even freezing days. But there are numerous indicators of the new season all around. Mornings have recently been filled with the "jenny jenny jenny" songs of Northern Cardinals, the mournful calls of Mourning Doves, and flocks of Cedar Waxwings have arrived from wherever they spent the winter. Ball Moss seeds have been flying for several weeks and clinging onto an assortment of objects, dead and alive. And a few butterflies and moths have already emerged from their cocoons; Sulphurs and Fritillaries are flying during the warmer afternoons.

The second week of February brought several other happenings in nature that might truly be considered harbingers of spring. Several Eastern Redbud trees began blooming; these bright purplish-red flowers appear several weeks before the first leaves. The first Purple Martins have been reported along the coast and at a few inland sites; adult males, sometimes called *scouts*, reach our area of South Texas first, and females and subadult birds arrive a few weeks later. And several of the earliest flowers are starting to grow: Texas Thistles are forming rosette patterns in fields and along ditches, Wild Onion leaves are sprouting in pastures and lawns, and Bluets already are flowering in moist, sandy areas.

Agarito buds are already full in my yard, and I know that any day now they will produce bright yellow flowers with a sweet scent that can be detected for ten feet or more. Honeybees and other nectar-lovers will magically appear to gather the fresh pollen. In a few more days the Huisache flowers will appear to fill the pastures with their dominance. Then it will be time for the abundance of Texas Bluebonnets, Indian Paintbrush, and dozens of other wildflowers.

All of the above herald spring, and dozens more are indicators of the annual awakening. The optimism of spring is everywhere. It is a wonderful time to be alive, to wander outdoors and experience nature at its best. Promise is all around us, and fulfillment is just ahead. Blossom by blossom, bird calls and songs, spring begins.

The Sharp-shinned Hawk Is Our Common "Bird Hawk"

FEBRUARY 28, 1999

Every once in a while one is fortunate to experience Mother Nature up close and personal. My latest observation was of a Sharp-shinned Hawk that captured its prey barely ten feet in front of me. It happened so fast, however, that it was over in less than two or three seconds, except for a few dislodged feathers that slowly floated down to the ground.

I was sitting in my kitchen, looking out of the window onto my backyard, where I have a number of feeders—seeds, peanut butter, and sugar water—as well as a constant water dripper that splashes into a birdbath. Almost 100 birds were present at the time, either at the feeders or nearby on the ground. They included 30 or 40 Chipping Sparrows, 15 to 20 American Goldfinches, 8 to 10 Northern Cardinals, 8 to 10 White-throated and a couple of Lincoln's Sparrows, 4 or 5 Inca Doves, 2 or 3 American Robins, and a lone Carolina Wren. A Buff-breasted Hummingbird had just rushed up for a drink and then dashed off to the dense brushy area beyond my yard.

Suddenly, a small hawk came hurtling down onto one of the feeders, grabbed a Chipping Sparrow, and in the same motion continued on across the yard and disappeared into the brush. Two tiny feathers floated downward. The majority of the other birds exploded into flight, dashing away with amazing speed. Two goldfinches and one robin froze in place, huddled down to make themselves small and hidden. They remained still for more than a minute before they resumed feeding.

The Sharp-shinned Hawk is a winter resident in South Texas; I had seen it several other times, and I was aware that it had targeted my yard, undoubtedly due to the available prey species present there throughout the daylight hours. And although I had seen it make a number of earlier passes, this was the first time that I was able to observe it actually make a successful catch.

Sharp-shins are among our smallest hawks, about the same size as the common American Kestrel that also is present through the winter months. Sharpies are not as colorful or obvious as kestrels, however, as they depend upon their dull plumage and stealthy behavior to survive. They are closely related to the larger Cooper's Hawk that also is present in South Texas in winter. Cooper's Hawks possess a rounded tail while the tail of Sharp-shinned is squared. Both are *accipiters*, or bird hawks, also characterized by short, rounded wings and a long tail, useful for maneuvering through vegetation. They rarely soar overhead like the much larger Red-tailed Hawk (a broad-winged hawk) but spend most of their time in trees or brush, usually perched where they can watch for a careless songbird. Cooper's Hawks are large enough to prey on killdeer-sized birds, while Sharp-shinned Hawks usually take sparrows, chickadees, titmice, and occasionally robin-sized birds.

All the winter-only hawks of South Texas leave at about the same time as the winter-only songbirds. They are all members of the bird community that typically includes a variety of species: seed-eaters, insect-eaters, fruit-eaters, as well as predators. Their long-term survival depends upon each finding adequate food, water, and shelter during the seasons of the year.

If we are lucky, we occasionally are able to obtain a first-hand perspective on what a bird's existence is all about.

March

White-winged Dove Populations Are on the Increase

MARCH 1, 1998

A note from Annette Sanford of Ganado claimed that the flock of 10 to 12 White-winged Doves "that have hung around my neighborhood for about three weeks" is the first seen in the forty-five years that she has lived there. Annette's experience has been repeated at several other locations within the Golden Crescent. These large, rather gregarious doves can become fairly common once they become established. And, like their smaller Inca Dove cousin, they swiftly learn to take advantage of handouts at bird feeders.

White-wings began to move into South Texas only since the 1980s. They are far more common to the south, especially in towns from Corpus Christi and Beeville southward and throughout the Rio Grande Valley. They once were so numerous in the Valley that hunters would compete to see who could kill the greatest numbers. But with the extensive clearing of the native brush in the Valley to make room for citrus orchards during the 1920s and 1930s, the White-winged Dove population declined drastically. They have since recovered to a considerable extent, once they began to nest on orchard trees and to frequent more urban habitats.

White-wings feed on a variety of plant materials, including all sorts of seeds, mast, and fruit, depending on the time of year. Year-round, doveweed and spurges, sunflowers, and cultivated crops, such as sorghum, ranked highest in volume and occurrence on a statewide study that was reported by Clarence Cottam and James Trefethen, in their 1968 book, *Whitewings: The Life History, Status, and Management of the White-winged Dove.*

The White-winged Dove is a large dove, about the size of the feral Rock Doves or pigeons that we find in the city parks and on our overpasses. White-wings are easily identified by their size, obvious white wing bar that is especially evident in flight, rounded tail with white edges, and marvelous song. The song is usually described as a drawn out, cooing call, like "who-cooks-for-who," that may be repeated time and again during spring and early summer. And they also possess a fairly distinct courtship behavior of flying over their breeding territory and soaring back with wings out to display the contrasting pattern.

The other common doves found in South Texas include the abundant Mourning Dove and two much smaller doves: the Inca Dove and Common Ground-Dove. The Inca Dove is commonplace in the towns throughout South Texas and at almost every developed site in the region; they can be especially numerous at feeders. Inca Doves show a scaly plumage and a reddish wing-patch, possess a long tail edged with white, and sing a double "cooo-coo" song. The Common Ground-Dove lacks the scaly back and has a short reddish bill tipped with black, a short, rounded tail without white edges, and sings a soft, ascending "wah-up" song. Ground-Doves prefer more open, drier sites and rarely are commonplace in yards and about feeders. Three other nonnative doves—Ringed Turtle-Dove, Eurasian Collared-Dove, and Spotted Dove—are also occasionally seen in South Texas but are not yet known to breed in the wild. If anyone finds a wild population of either of the three latter species, I would be very interested in learning about them; these three exotics seem to be on the increase in the eastern portion of the state.

Three other native doves/pigeons occur in Texas: White-tipped Doves are common along the Rio Grande and range as far north as Refugio County; Red-billed Pigeons are summer residents along the Rio Grande below Falcon Dam; Band-tailed Pigeons are resident in the West Texas mountains. And two Mexican species have been recorded in the southern portion of the state a few times: Ruddy Ground-Dove and Ruddy Quail-Dove.

It is fairly obvious that Texas is blessed with a good variety of columbids—pigeons and doves that are members of the *Columbidae* family. So an obvious question might be asked about the difference between a pigeon and a dove. It's in the name only and not worthy of much discussion. I have often told those individuals who ask that a dove is only a small pigeon and a pigeon is only a large dove. And the most handsome of all these is the White-winged Dove.

Cliff Swallows build nests of mud pellets.

Cliff Swallows Are Returning to South Texas

MARCH 2, 1997

One sure sign of spring is the arrival of Cliff Swallows. These little birds build their mud-pellet nests under highway bridges and culverts and on barns and similar places throughout South Texas.

Colonies of from 35 to more than 200 individuals usually can be found at all the area's concrete bridges and overpasses. They arrive in South Texas by early March and leave their nesting sites by the end of August. A few migrants can usually be found through October. Our Cliff Swallows spend their winter months in South America.

Nest-building is an amazing activity. Although some Cliff Swallows may only refurbish an old nest, most begin anew by constructing a retort-shaped structure (tubular entrance to a spherical cavity) from thousands of tiny mud pellets that they paste together literally one at a time. They congregate at mud puddles or along the banks of streams to gather mud that they shape into round pellets in their beaks. They then methodically construct their nests. Construction time lasts for about five to fourteen days, depending upon the availability of mud and adequate food. They then line their nests with grass and feathers, and the female lays four to five spotted eggs.

Biologists have discovered that Cliff Swallows practice "intraspecific brood parasitism," by laying eggs in nests other their own. And, surprisingly, some individuals can transport their eggs to another nest. They may even toss out an egg, presumably to replace it with their own. As many as 25% of all Cliff Swallow nests in a colony may be parasitized.

One of the square-tailed swallows, in comparison with the long-tailed Barn Swallow and fork-tailed Tree, Bank, and Rough-winged Swallows, Cliff Swallows possess a buff-colored rump and cheeks, pale forehead, and blackish throat and back. They are most closely related to Cave Swallows, which we also have in South Texas. While Cliff Swallows nest in open places, Cave Swallows build their nests in twilight sites, such as in caves, culverts, and deserted structures.

All swallows are insect-eaters, taking millions of flying insects daily. Thirty-five Cliff Swallows collected in the vicinity of cotton fields in Texas had consumed 687 boll weevils, averaging 19 in each bird's stomach. Beetles of all types are readily consumed. Other food types include ants, bees and wasps, flies, and a number of true bugs. Various small fruits are also eaten after the nesting season.

Cliff Swallows are wonderful neighbors for a number of reasons. Not only do they eliminate many of our insect pests, but they provide one more reason to admire and wonder about our natural environment.

What's a Bird Worth?

MARCH 8, 1998

With spring already happening in South Texas, thousands of birders are flocking to the state. These visitors will spend more than $155 million in 1998, according to the American Bird Conservancy.

Income generated by the birding industry in Texas and throughout the country has exploded during the 1990s. The 1994–1995 National Recreation Survey revealed that the hobby of birding increased by 155% in ten years, while golf increased only 29% and fishing, hunting, and tennis actually declined by 4%, 12%, and 29%, respectively. The survey also claims that 191,000 jobs, including 4,730 in Texas, were contributed by nonconsumptive bird use. Nonconsumptive bird recreation activities generate $516 million in federal income taxes, $306 million in state taxes, and $73 million in state income taxes.

In Texas, birders annually spend nearly $20 million while visiting Laguna Atascosa National Wildlife Refuge; the 99,000 birders visiting Santa Ana National Wildlife Refuge spent approximately $34.5 million, including $14.4 million in nearby communities. More than 70% of those birders are visitors from out of the state.

Birding will undoubtedly only become more popular in Texas, what with the added birding opportunities available. The first-of-its-kind Great Texas Coastal Birding Trail, for example, has already increased visitation to communities along the central Texas coast since its inception in 1997. And twenty-four of the central trail's ninety-four sites are located within Victoria, Calhoun, Jackson, Refugio, Goliad, and DeWitt counties.

Texas also has initiated a number of birding festivals, more than any other state, ranging from one- to three-day affairs. These include such wide-ranging activities as the Eagle Fest in Emory, CraneFest in Big Spring, Whooping Crane Festival in Port Aransas, Attwater's Prairie Chicken Festival in Eagle Lake, Bluebird Festival in Wills Point, Migration Celebration in Brazoria County, Rockport's Hummer/Bird Celebration, and Rio Grande Valley Birding Festival in Harlingen. The latter festival attracted one thousand folks its first year and had an economic impact of $266,000 for the three-day event; more than eighteen hundred people attended the 1997 festival, providing $1.6 million to the local economy.

In addition to all that, Texas is the number one bird state, with more species recorded than any other state, and the Christmas Bird Counts taken in Texas consistently rank the highest in the nation. The point is that Texas, more than any other state, has tremendous advantages year-round for birders and other nature enthusiasts. So it seems to me that more needs to be done in the Golden Crescent to take advantage of these interests.

What can be done? Maybe the area chambers of commerce should establish a combined task force to look into what opportunities are available, with the idea of developing certain sites for birding. Although a few projects have already been initiated, such as the Athey Nature Trail in Victoria's Riverside Park and Dupont's eagle tower and wetland project (located in east Victoria County), these are not enough to make birders stay for more than a few hours. Birders already are passing through South Texas and need only some additional incentive to stay longer and enjoy the abundant birding opportunities that those of us who live here already appreciate. Birds are big business!

Texas Paintbrush Is an Early Bloomer

MARCH 9, 1997

An abundance of Texas Paintbrush blooming along the roadsides is synonymous with spring. One of the earliest spring flowers, its red bracts often are mixed with the deep purples of Texas Bluebonnets. This mixture usually remains until the growing grasses overcome the brighter reds and purples. But the Texas Paintbrush also will appear in fewer numbers throughout the summer and fall months.

Our springtime Texas Paintbrush, sometimes known as Indian Paintbrush, Indian Pink, or Painted Cup, is but one of nine species of paintbrushes that occur in Texas. But *Castilleja indivisa,* as it is know to scientists, is the only one that blooms in spring in South Texas. And it is the only one that is an annual, lasting a single year—rarely a second year.

The other eight species of paintbrushes are perennials, persisting for several years. Their common name is derived from its appearance, like a ragged brush dipped in paint. But unlike most flowering plants, the "flowers" of paintbrushes are like those of poinsettia in that the red portion is really composed of leafy bracts, or modified leaves. The true flower (petals and corolla) is less obvious and attractive.

Most of the native paintbrushes were utilized by Native Americans in several ways. Although it is rather toxic, a weak tea was made from the flower to treat rheumatism, to purify the blood, and to soothe burned skin and insect bites. And dyes were made from the plants as well: black from the roots and greenish yellow from the stems, leaves, and flowers.

Another interesting fact about paintbrushes concerns their relationship with other plants. They are *hemiparasitic,* in that they actually obtain water and minerals from a host plant. They are more vigorous when they are in contact with another plant, such as the Texas Bluebonnet. Experiments done at the Lady Bird Johnson Wildflower Research Center in Austin suggest that in times of drought paintbrushes obtain so much water and minerals that the host plants often wilt and die more rapidly.

Each year the Texas Department of Transportation sows the roadsides with thousands of pounds of Texas Paintbrush seeds. Since there are well over 5 million seeds to a pound, that amounts to lots and lots of paintbrush seeds. Is it any wonder we have such colorful spring roadsides?

Blackbrush Acacia Is in Full Bloom

MARCH 12, 1999

Blackbrush Acacia, sometimes called "Chaparro Prieto," is currently in full bloom throughout much of South Texas. This thorny little tree or shrub possesses white or light yellow flowers that appear in cylindrical spikes; they can be so abundant that they literally dominate the branch. The leaves are alternate and compound, with two to four (rarely five) pairs of leaflets, and they possess straight spines.

Most people, as well as bees, appreciate these lovely shrubs when flowering; they make especially good honey. However, like the closely related Huisache trees, with their bright yellow and scented flowers, they can take over a field and be a real problem to ranchers.

Four additional species of acacias, besides Blackbrush and Huisache, occur within the Texas Coastal Bend region: Catclaw (*Acacia greggii*) also possesses creamy yellow flowers that appear in cylindrical spikes, although it normally blooms from April to October. And Catclaw spines are decurved, not straight like Blackbrush. Fern Acacia (*Acacia angustissima*) and Guajillo (*Acacia berlandieri*) both possess white to cream flowers that are rounded instead of cylindrical. Guajillo can flower most of the year, from February to December, while Fern Acacia flowers from April to October. Finally, Twisted Acacia or Huisachillo (*Acacia schaffneri*) flowers are orange-yellow in color, very much like the common Huisache, and bloom from February to April. They can be most common after spring rains. This last acacia differs from Huisache by its spreading growth (not tree-shaped) and narrower and longer seed pods.

Acacias, all members of the legume or pea family, are widely distributed in warm regions around the world, where there are about 600 species; 300 occur in Australia. Ten species are native to Texas, and all of the Texas acacias flower in the spring and occasionally in the summer following rain. The genus *Acacia* means hard, sharp point, in reference to the prominent spines.

The legume family, or *Leguminosae*, is extremely large, with many representatives worldwide. Others in Texas include mesquite, mimosa, redbud, senna, retama, Mountain Laurel, and even clover and bluebonnets. Acacias and mimosas are often so similar they are difficult to tell apart. But the key difference is in the flowers themselves: acacia flowers possess numerous (twenty to one hundred) stamens, while mimosas possess ten or fewer stamens per flower. And mimosa fruits (pods) usually are flattened and sometimes contorted.

All these shrubs and trees are lovely when in flower!

Scissor-tailed Flycatchers Are Arriving in South Texas

MARCH 15, 1998

Of all the neotropical migrants that pass through South Texas, the lovely Scissor-tailed Flycatcher is probably the most welcome of all.

Few birds have the appeal of this charismatic flycatcher. Not only is it one of our most beautiful and gregarious birds, but it seems to prefer a relationship with humans, nesting on utility poles and in trees often surprisingly close to our various structures. Its amazing courtship flights and continuous singing tend to give it an additional appeal. It therefore is often called the "Texas bird of paradise." And its arrival in South Texas is a sure sign that the new season has begun.

The long-tailed, brighter males arrive first with the shorter-tailed females appearing a few days later. By then the males have already established territories and are chasing competitors away from preferred sites. When the ladies arrive, the males take on a very different persona, performing some wonderful courtship flights, ascending to more than 100 feet before sailing back, often with outstanding aerobatics. These dramatic flights include up and down flying, much zigzagging, and even reverse somersaults, usually at great speeds and with tail flowing and fluttering and wings out to display their salmon-colored armpits (axillaries) and underwing linings. All the while he is performing, a male flycatcher will be giving cackling-snapping calls. The female will usually join in the fun. Scissor-tails also give a unique dawn song on their breeding grounds that includes a series of loud stuttered "pup" notes that conclude with an emphatic "perlep" or "peroo."

Like all flycatchers, the Scissor-tail's diet is principally insects, at least during their nesting season. Although most insects are captured in flight,

Scissor-tails will also take insects on the ground, perhaps more often than most flycatchers. Grasshoppers are a favorite food source. After nesting and while inhabiting their wintering grounds, however, they will also consume berries when available.

Although paired Scissor-tails are generally loners, as soon as the youngsters are fledged, they will usually join other family groups. In some cases these flocks can include up to 200 individuals. And unlike most other members of the flycatcher family, which usually are quiet after nesting, Scissor-tails continue calling until they leave for their wintering grounds in September or October, as well as throughout their migration and in winter. These flocks often congregate at choice sites. And 100 or more Scissor-tailed Flycatchers can create quite a racket.

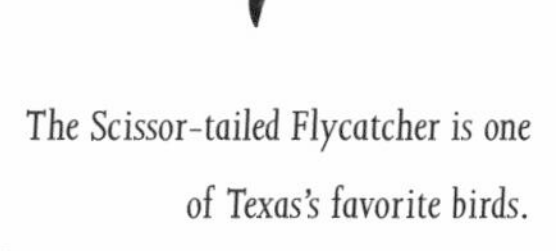

The Scissor-tailed Flycatcher is one of Texas's favorite birds.

Many Texans think of this bird as their state bird instead of the Northern Mockingbird, which is the official state bird. That undoubtedly is because of the charisma of this long-tailed songbird, and also perhaps because the mockingbird is so commonplace. While mockingbirds are full-time residents throughout most of the state, leaving only the far northern portions of the state in winter, Scissor-tailed Flycatchers normally are present only from March through October. But during that period, they can be found in all but far West Texas, where they occur only occasionally.

By November the vast majority of the summer residents and migrants passing through the state from Oklahoma, Kansas, and southeastern New Mexico have gone south. Recent records, however, suggest that lone birds may remain in South Texas all winter. The rest migrate south to central Mexico and into Panama, where they occur in huge flocks, utilizing open grasslands, pastures, and fields.

But by March they are with us again. Few songbirds are as well loved and admired as our lovely Scissor-tailed Flycatcher.

Cedar Waxwings Are Regular Spring Visitors in South Texas

MARCH 16, 1997

Every year about mid-March, when spring flowers are starting to appear and the first of the northbound migrants are being reported in South Texas, flocks of Cedar Waxwings appear.

Some of these lovely waxwings are surely neotropical migrants, wintering in the Tropics, but others may have remained north of the border. Now they are gathering in larger flocks and feeding on last year's berries and the new buds that are developing on our trees and shrubs. Mulberry, cedar, Mountain Ash, and Pyracantha berries are favorites. Choice feeding sites often become staging areas for gathering flocks of migrants.

Cedar Waxwings are one of our best known birds. They are easily recognized by their size (7–8 inches long), distinct, pointed topknot, black mask against a soft brown head, pale yellow belly, and a bright yellow tip on their otherwise gray and black tail. And their soft, high-pitched, slightly trilled whistle also is distinct. A wheeling flock of Cedar Waxwings, flying in unison, can hardly be mistaken.

Cedar Waxwings are truly a social species. These gregarious creatures rarely are alone, but flocks of a few to 100 or more are typical. Even on their nesting grounds, only in the Texas Panhandle and northward, many individuals will share common feeding areas. Although fruit and buds are utilized most often in South Texas, they also will feed on a wide variety of insects, oftentimes taken in mid-air, like a swallow or flycatcher. They may even alight on the ground to feed, and they also come to the ground to drink. I have often found a ring of 8 to 10 individuals on my birdbath. Like American Robins, waxwings seem to drink a great deal of water.

They also posses some rather unusual habits. Harry Oberholser in *The Bird Life of Texas* tells about a "charming ritual" in which a pair or group of Cedar Waxwings, sitting together side-by-side on a branch, will "pass a cherry back and forth before one swallows it," or, in courtship, a pair may pass flower petals or insects back and forth.

But it is their gregarious feeding habit that is best known. Hundreds of waxwings, that descend upon a tree or shrub filled with berries or buds, can strip the food in a few minutes. And in spite of their normally quiet and dignified manner, they can become avian clowns after feeding on fermented berries. There are numerous reports of drunken waxwings falling off branches and hanging upside down while attempting to right themselves. They may fall to the ground and put on a grand show of attempting to stand upright, teetering about or running in circles. But eventually they recover and continue their journey toward their breeding grounds.

In South Texas, we normally can enjoy these lovely creatures only during spring.

Scientists Slowly Solving Avian Navigation Puzzle

MARCH 17, 1996

At this time of year, millions of birds of all sizes and kinds are passing through South Texas. The majority of these are neotropical migrants.

Many of these northbound birds fly across the Gulf of Mexico, leaving Mexico's Yucatán Peninsula during the early evening hours. They normally arrive along the Gulf Coast the following day between mid-morning and early afternoon. Others follow the Gulf Coast northward, passing directly through South Texas.

In the big picture of migration, it is understandable that these spring migrants head north toward their nesting grounds. But what is less clear is how they are able to navigate from one tiny speck of habitat in the Tropics to another speck of habitat that may be as much as 4,000 miles to the north. How do they manage to find the exact same place at Riverside Park, on the shore of Lake Michigan, or on the Alaska tundra where they were fledged?

The answer to that question is not fully understood, but some pieces of the avian navigation puzzle have been found. For instance, it is well known that birds are fully capable of following principle landforms, such as riverways, seashores, and mountain ranges. This suggests, of course, that once a bird has traveled the route once, even in the opposite direction, it can do so again. Domestic pigeons released miles from their roosts fly in circles until they are able to recognize their home surroundings. But what about nighttime migrants or those birds flying over fog or clouds?

Birds also are able to use celestial navigation; they can navigate by the stars. Biologists have placed spring migrants in a circular cage in a planetarium, and when stars came out, the birds began hitting the north walls. When the northern stars were reversed, the birds began hitting the south walls.

To take this experiment even further, spring migrants (equipped with tiny radio transmitters) were placed in a cardboard box that was suspended from a helium-filled balloon. When the box was opened from aloft on a clear night, the birds flew erratically for only a minute or two and then set out on a straight northward course. If the birds were released on an overcast evening, when an afterglow was still visible in the west, they were able to orient themselves by that method. But when neither clue, stars or an afterglow, was available, they flew downwind, even if the wind was blowing in the direction opposite their destination. When that direction proved to be incorrect, they eventually had to turn back.

Migrants often remain in choice feeding areas, awaiting the right cue to head out. They make a go or no-go decision based on what the weather will be like over the next several hours. They apparently are able to detect the most reliable conditions, when they will be aided by a tailwind and not encounter severe storms along their route. They are correct the majority of times.

Upland Sandpipers Making Way through South Texas

MARCH 19, 1995

The mellow calls of Upland Sandpipers, like a mournful "hoo-lee" or "hoo-lee-lee," are beginning to echo across the open countryside of South Texas.

These tall, stately shorebirds are already en route through South Texas on their northbound migration between their wintering grounds in southern South America and their breeding grounds on the Great Plains and north to the tundra of central Alaska. There they will select a territory, court and mate, raise a family, and be back in South Texas on their southbound journeys by August.

Few shorebirds possess the charisma of the Upland Sandpiper. Its upright stance, wonderful mellow songs, loud, ringing alarm calls, and its habit of sitting on high posts make it especially appealing. And on its nesting grounds, it will perform amazing distraction displays to tempt the intruder away from its nest. Males will often drag a wing, pretending an injury, running away and stumbling as if it were easy prey. But just as a predator is about to capture the injured bird, it will recover and fly off with loud "quip-ip-ip" notes. It will then perch on the very top of a tree or shrub and chastise the intruder for as long as it remains.

It seemed so strange last summer to find this bird, that passes through our South Texas area in spring and fall, defending a piece of Alaska tundra. But once I recovered from its presence and its incessant calls and wing-dragging behavior, I realized that its breeding grounds did not look all that different from our coastal plains. Although our coastal grasslands lack permafrost, the exterior appearance and the great numbers of mosquitoes are similar.

Upland Sandpipers are not the only shorebird that frequents the open grasslands of South Texas in spring. The American Golden-Plover, a chunky bird with a mottled, golden back, black face and underparts, and a white forehead and sides, is also present in substantial numbers. Golden-Plovers pass through South Texas somewhat earlier than Upland Sandpipers; the Golden-Plovers are en route to their breeding grounds north of the Arctic Circle. Their call notes are loud whistled "kleep" or "queedle" notes.

This bird has one of the longest migration routes on record. These Arctic nesters also spend their winters in southern South America. Southbound birds fly east to eastern Canada where they cross the Atlantic Ocean between Nova Scotia and Venezuela in one 2,000-mile flight over water, then continue south to Argentina for the winter. Northbound birds follow a completely different route that takes them through Central America, northward along the Gulf Coast of Mexico and Texas, and then straight north into northern Canada and the Arctic. Their northbound migration is approximately 8,000 miles in length.

Although Golden-Plovers pass through South Texas earlier than Upland Sandpipers, the species can usually be found together, foraging on fields and pastures during late March and early April.

Spring's Early Butterflies Are Taking Wing

MARCH 24, 1996

Even before our earliest migrant birds began to move through South Texas, a number of butterflies were on the wing. Although some of these lovely creatures may have survived the winter in protected places, the majority of our springtime butterflies emerged from their chrysalis during our warmer days. And now those early butterflies are taking advantage of the early flowering shrubs.

In my Mission Oaks yard near Mission Valley, I identified 17 butterfly species in February and early March: Pipevine Swallowtail, Cloudless Sulphur, Southern Dogface, Little Yellow, Gray Hairstreak, Gulf Fritillary, Texan Crescent, Painted Lady, Red Admiral, Common Buckeye, White-striped Longtail, Northern Cloudywing, Mournful and Funereal Duskywings, Common Checkered-Skipper, and Fiery Skipper. The most numerous and obvious of these were the Pipevine Swallowtail, sometimes called "Blue Swallowtail," and the lemon-yellow Cloudless Sulphur.

It is next to impossible to miss the Pipevine Swallowtail because of its large size, more than 3 inches from wingtip to wingtip, and its black-and-blue pattern. When resting, it also shows a series of red to orange and black spots on the margin of the underside of its hind wing. This extremely attractive swallowtail occurs throughout North America, and it favors honeysuckle, milkweed, azalea, lilac, and thistle flowers.

The Cloudless Sulphur is a smaller, all lemon-yellow butterfly with a wingspan of about 2½ inches, and it usually is seen hurrying by as if it is en route to some distant location. However, if you watch carefully, it will often return to land briefly on a flower to sip the nectar. On a closer look, you can often see reddish brown mottling or dots below. The very similar Southern Dogface usually is greenish and shows faint pinkish stripes on the bottom of its forewings.

Of all the spring butterflies so far, however, the Gulf Fritillary is the brightest. It sports brilliant orange-red wings with black dots and streaks on the upper side and black-and-silver slashes and dots against a dull to rich brown background on the under side. This is an extremely nervous butterfly that is difficult to see well, but the patient observer will be well rewarded. Its name comes from the fact that it haunts the Gulf of Mexico region and often is found flying far out over the water. This lovely creature uses Passion Flowers as its host.

Which is the best butterfly identification guide? Although none include all the species we are likely to see in South Texas, the *National Audubon Society Field Guide to North American Butterflies* is probably the best. Butterfly names were standardized by the North American Butterfly Association and published in a 1995 checklist by that organization.

Learning Birdsongs through Mnemonics

MARCH 30, 1997

How many birds can you identify by their songs alone? Most folks know more birdsongs than they might think. For example, who doesn't recognize the song of a Bobwhite Quail? How about a Blue Jay? Cardinal? American Crow? Chickadee? Maybe even Barred or Great Horned Owls, Eastern Screech-Owl, Carolina Wren, and Common Yellowthroat? See, you know more birdsongs than you first thought.

Bird identification is just as sure by sound as it is by sight. In fact, many species are so difficult to see in the high foliage that, especially for biologists doing populations studies, birdsongs are used for identification more often than the actual sightings. Birders actually utilize the bird's phonetic rendering. For instance the mnemonic used to describe a common song of the bobwhite is "bob-white." Blue Jays sing "jay, jay, jay." American Robins, one of our most cherished birds, sings "cheerily-cheery-cheerily-cheery." Carolina Chickadees sing a whistled "fee-bee, fee-bu." Tufted Titmice whistle "peter, peter, peter." And one of my favorite song is that of the Carolina Wren: "teakettle, teakettle, teakettle." And how about the cardinal? It sings "what, cheer, cheer, cheer, cheer, cheer, cheer, whot, whot, whot," or "birdy, birdy, birdy."

The use of mnemonics is most useful when walking through the woods that echo with the songs of a dozen or more neotropical songbirds. White-eyed Vireos sing "quick-with-the-beer-check." Red-eyed Vireos sing a continuous song, like "look up . . . see me? . . . over here . . . higher" and so on. Blue-gray Gnatcatchers offer a lispy "spee, spee, spee." Eastern Wood-Pewees sing a plaintive whistled "pee-ah-wee, pee-err." While the Common Yellowthroat, numerous in our wetlands, sings an easy-to-remember song, "witchy, witchy, witchy, witch."

The vast majority of South Texas birds possess a song, although fewer than half of the almost 9,500 known bird species actually sing. But many species possess a repertoire of songs, often singing different songs in order, one after the other. Our Northern Mockingbird has as many as 150 songs, while a Brown Thrasher can sing more than 3,000 song types. A European Starling's repertoire may include as many as 67 song types. And many wrens, especially the tropical wrens, often sing duets, so that one individual begins the song and its mate ends the song.

How many songs do birds sing in a single day? That varies with the species. Ornithologist Margaret Nice recorded 2,305 songs in a single May day for a Song Sparrow. She also reported a Black-throated Green Warbler that sang 1,680 songs in seven hours, and she estimated that on a typical day of sixteen daylight hours, he would have sung more than 3,000 songs. But the North American winner is the Red-eyed Vireo. Ornithologist Harold Mayfield recorded a Michigan Red-eyed which sang 22,197 songs in a day.

Biologists tell us that birdsongs are utilized to identify the bird's territory, usually directed at other males, and to attract a mate. The song may also serve to convey a message. But whatever their purpose, most listeners appreciate birdsongs simply for their acoustical quality. For many of us, it would be an empty world without the songs of birds.

Skunks Are on the Trail of Romance in Spring

MARCH 31, 1996

In spring, like so many other critters, including humans, skunks turn their attention to love. They begin to wander about in search of that one true love of their life or, more truthfully for skunks, their annual affair.

Skunks normally live a solitary existence, only pairing up during their spring breeding season. Consequently, they are far more likely to meet their demise from high-speed vehicles in early spring, usually in February or March, when they are searching for a mate, than at any other time of the year.

The South Texas region has three species of skunks: Striped, Spotted, and

Hognose, but only the Striped Skunk is reported regularly. The smaller and less common Spotted Skunk, identified by numerous white markings, is far more secretive and is rarely seen. And the larger Hognose Skunk, with a longer snout and an all-white back and tail, is more numerous in the south and only rarely is reported in the Coastal Bend.

The Striped Skunk is easily identified by its black body with narrow white stripes that runs from the top of its head backward along its back, like an elongated V. About the size of a large house cat, Striped Skunks can appear almost anywhere, from our fields and woodlands to even our residential areas. Nocturnal in behavior and rarely encountered, they are more often detected by scent than they are seen.

All skunks possess scent glands with an obnoxious odor that they can spray at an antagonist when disturbed. Although it may seem that the typical skunk odor is commonplace, they spray only as a last resort. The scent glands, located near the base of the tail, are normally activated only after the animal warns the intruder first. It first will audibly strike the ground with its forefeet and even make short rushes at its enemy before actually using its potent spray. It finally will bring its rear around toward its enemy, with its tail erect, and then discharge fine yellow droplets through small ducts that open just inside the anus. These glands are encased in muscles that can be

Striped Skunks can be common in South Texas.

voluntarily controlled by the animal when the situation demands it. The powerful scent may be detected miles away during favorable weather.

Although skunks are usually considered bad neighbors, due to their odor and an occasional invasion of chicken coops, they normally are good friends to farmers and ranchers. Typical skunk food includes grasshoppers, grubs, beetles, snakes, frogs, rodents, crabs, and an occasional bird and egg.

The skunk's few enemies consist of humans, large dogs, and Great Horned Owls, the only nighttime predators large enough and aggressive enough to kill a skunk. The skunk's powerful defense immunizes them effectively from most potential enemies.

April

Aggressive Cattle Egrets Are Spring Arrivals in Texas

APRIL 2, 1995

Three all-white egrets frequent South Texas year-round: the tall Great Egret, with a heavy yellow bill and blackish legs; the smaller, trimmer Snowy Egret, with a black bill and legs and yellow feet; and the stocky Cattle Egret, with an orangish bill and legs. And during the breeding season the Cattle Egret also is adorned by orange-buff patches on its crown, chest, and back.

The Cattle Egret is distinct because it is the only one of the three that spends its time in fields, often associated with cattle, while the Great and Snowy Egrets almost always are found at ponds, streams, or wetlands. While the Great and Snowy Egrets feed on fish, frogs, and other aquatic creatures, Cattle Egrets feed principally on insects and other invertebrates that they find in the fields and on cattle.

The Great and Snowy Egrets have been part of the South Texas environment for longer than anyone can remember, but the Cattle Egret did not arrive in Texas until the 1950s. Since then it has increased in numbers and distribution, and today it can be found everywhere in Texas except in the highest mountains in West Texas. And unlike the Great and Snowy Egrets that rarely are found in large numbers (except in their roosting and nesting sites), Cattle Egrets are almost always found in congregations of a few to 100 birds or more.

Perhaps that gregarious behavior helped with its incomparable invasions of South, Central, and North America. Cattle Egrets are native to the vast savannah environment of Africa, where they feed alongside the native graz-

Cattle Egrets, unlike other egrets and herons, find their food in fields and pastures.

ers, such as antelopes, zebras, and elephants. They often perch on the backs of these "cattle" to feed on ticks and flies. This behavior provided them with the names of "cattle" or "tick" birds.

However, sometime during the 1930s, they suddenly appeared in South America, probably the result of a hurricane that carried a few across the Atlantic from West Africa. The South American populations increased dramatically, and within twenty years they began to appear in South Florida. The first Texas record, according to Harry Oberholser's *The Bird Life of Texas,* was an immature bird found on Mustang Island in November 1955. The first Texas nesting records were reported for Galveston in 1958, and by October 1967, the Cattle Egret was recorded at Big Bend National Park in West Texas.

The aggressive Cattle Egret is somewhat of a migrant in South Texas, although a few can usually be found at preferred sites year-round. But the majority of our birds move south into Mexico for the winter months, returning to our fields and pastures in spring. Flocks of a few to several dozen can be often found at this time of year, flying rather low in groups or trailing out in scattered flocks, heading northward. At night, they all congregate at favorite roosting sites, usually along a river or near water, where they may also nest in spring and early summer, but at dawn, they return to the fields to feed. And like their ancestors long ago on the African savannahs, they can often be found feeding at the feet of cattle or perching on top, searching for ticks and insects.

Although the Cattle Egret may not be a pure Texan like its two all white cousins, it is now a significant and valuable member of the avian community.

How to Attract and Keep Hummingbirds

APRIL 5, 1998

Ruby-throated and Buff-bellied Hummingbirds have already arrived in the Coastal Bend, where they will remain throughout the summer months. Thousands of other Ruby-throats are streaming northward. Those passing hummers will spread out and nest all across the eastern half of North America, from central Texas northeast to the Maritime Provinces and west to Alberta, Canada.

The hummingbirds that remain with us can usually be encouraged to feed at artificial feeders that can be placed almost anywhere in the outdoors. Locating a feeder just outside a window, so that one can obtain good close observations, works just as well as those located at a distance. Hummers seem to have the ability to find almost any feeder in the open. And they will utilize our feeders all during their stay so long as the feeders are kept clean and are filled on a regular basis.

Sugar water, at one part sugar to six to ten parts water, is the not-so-secret ingredient. Red food coloring is not only unnecessary but may be harmful to the hummers. So long as the feeders contain a red or yellow entry place, that is all that is required. Commercial mixes are also available; they are not only expensive but totally unnecessary. Some commercial mixes contain vitamins, which sounds like a good idea initially, but since only a small part of a hummingbird's diet consists of sugar water, special vitamin mixes are a waste of money. Plus, added nutrients in the feeders encourage pesky microorganisms that may be harmful. A hummingbird diet naturally includes approximately 65% insect life that supplies all the nutrients needed by a healthy hummingbird.

The very best way to good hummingbird health is to encourage uncontaminated populations of insects in your yard. Avoiding the use of pesticides and encouraging your neighbors to do the same is most important. Second, you can add plants to your yard that provide food for aphids and other small insects. Or you can cultivate fruit flies by placing overripe fruit in warm and accessible sites. A pinch of baker's yeast will start fermentation.

Finally, plant shrubs, vines, and flowers that provide nectar. Living in South Texas, we have numerous options. But since I have been asked on several occasions which plants are best, here is a list that I recommend (alphabetically, not in order of importance): Beebalm, Belize Sage, Butterfly Bush, Cape Honeysuckle, Cardinal Flower, Coral Honeysuckle, Crossvine, Cypressvine, Firebush, Firecracker Vine, Firespike, Flame Acanthus, Fountain Flower, Giant Turk's Cap, Grapefruit-Scented Sage, Jewelweed, Mexican Bush Sage, Mexican Butterfly Milkweed, Mexican Cigar Plant, Mexican Fushia Sage, Mexican Sage, Mexican Sunflower, Pineapple Sage, Red Buckeye, Rosebud Sage, Shrimp Plant, Scarlet Sage, Smuggler's Vervain, Sweet Olive, Tree Tobacco, Trumpet Vine, Turk's Cap, Vanhouitte Sage, and Winter Honeysuckle.

Any yard that contains all or the majority of the above species is likely to accommodate hummingbirds year-round. Although some of these plants flower in spring, while others flower in summer and/or fall, a few are likely to flower during most winters. Ruby-throated, Black-chinned, and Buff-bellied Hummingbirds will thank you from March to October, and Rufous, Broad-tailed, and Anna's may move in during the winter months. Any hummer is worth a little extra time and money.

Hummingbirds Are among the Earliest of Our Neotropical Migrants

APRIL 5, 1996

My first Ruby-throated Hummingbird, a brightly marked male, arrived at my feeders on March 12 last year. Its arrival may have provided the last wintering Rufous Hummingbird with whatever signal it needed to move out and head for its breeding grounds to the northwest. It left a day or two later. And within a few more days, several more Ruby-throats began to take advantage of my feeders.

A neotropical migrant, Ruby-throats are among the first birds to begin the northward migration, usually reaching South Texas from late February to mid-March. Arrival dates depend upon weather conditions, and 1995's later-than-normal Ruby-throat arrival can undoubtedly be explained by the very odd weather we experienced. This eastern species can be expected in Arkansas by late March, in Illinois by mid-April, and in Chicago and New York City by the first week of May.

Although the larger Buff-bellied Hummingbird also spends the summer months in South Texas, the Ruby-throat is far more numerous and often is the only hummingbird species we may see. Adult males possess the distinct ruby-colored throat, emerald green head and back, all black bill, blackish tail, and whitish underparts. Less than four inches long, Ruby-throats are truly amazing creatures. They are able to make vertical takeoffs, hover, pivot on a stationary axis, and fly forward, backward, sideways, and upside down, with their wings beating about eighty times per second in a forward flight and two hundred times per second while performing display dives.

As might be expected by this behavior, hummingbirds have the most rapid metabolism of all birds. They consume the greatest amount of food relative to body weight of any vertebrate. They must feed almost continuously to remain alive. If we had the same metabolic rate as hummingbirds, our daily intake of food would be twice our body weight, our body temperature would be over 750 degrees, and we would use 155,000 calories a day.

Approximately 85% of a hummingbird's feeding time is spent foraging for nectar, an extremely high energy source. Flower nectar is composed of various sugars and water which are digested easily and quickly. The remaining 15% of a hummingbird's feeding time is spent looking for and consuming tiny insects and spiders that provide essential proteins, fats, minerals, vitamins, and roughage. Insects are captured in flight, on flowers, or even stolen off spiderwebs.

Ruby-throats are easily attracted by plantings and hummingbird feeders filled with sugar water. To make your own sugar water, simply mix one part sugar in six to ten parts water and bring to a boil. Let it cool before filling the feeder; the rest can be stored in the refrigerator. During hot weather, the sugar water must be changed at least every four to five days. And the feeder must always be kept clean.

The use of hummingbird feeders is a sure way to keep these amazing creatures around. Their antics provide us with many hours of enjoyment.

Birds Preen Plumage to Pure Perfection

APRIL 12, 1998

Have you wondered how it is possible that the white feathers of an egret or ibis remain so clean? This is in spite of the fact that many of these birds frequent muddy ditches and mudflats, where their plumage can easily be soiled. And yet, there they are, hunting their prey in these locations, without even a spot of dirt. What does it take to keep those immaculate feathers forever white?

The truth of the matter is that those feathers do become spotted and dirty, but each species takes meticulous care of its plumage. Each feather, one by one, is cleaned by the bird. This is done by the bird grasping each feather at the base with its bill, and then passing the feather through its partly clamped bill, nibbling its way toward the tip, removing oil, dirt, and ectoparasites. Or it may simply pull the feather through its partly clamped bill in one movement. This activity also smoothes the feather-barbs so that they lock together more tightly. Most birds also oil their feathers by preening, the act of taking oil from a gland near the base of the tail and working it into the feathers. This is especially important for waterbirds, but oil from the preen gland also helps keep the feathers clean.

Bathing is another important habit of birds in keeping their feathers clean and healthy, and some species, such as robins, bathe more often than other species. Preening is most important after bathing, as the feathers usually need another protective coating. And some individuals take more time preening their feathers than others. During this process, a bird will usually fluff out its body feathers to get at them more easily. The bird often bends and twists its neck to reach less accessible locations, and it even preens its head feathers by scratching it or by rubbing it against other parts of its body.

Some species, including egrets and herons, parrots, pigeons, and some hawks and owls, practice mutual preening. Usually done by paired birds, it requires each to alternately preen the other, especially on the head and upper neck. This mutual caressing is considered a form of pair bonding.

Another closely related feather-cleaning activity undertaken by a wide variety of birds is anting, the act of taking an ant or another object in its bill and rubbing the ant over its feathers. Some birds may actually sit on an ant hill and let the ants swarm over its body. The ants actually gather parasites from the bird's body and feathers. The bird may also take an ant, usually those with a high level of formic acid, and rub the ant juice onto its feathers. The acid tends to keep the feathers free of ectoparasites.

And some birds will substitute other invertebrates and other objects in their act of anting. Beetles, mealworms, pieces of lemon peel, cigarette butts, coffee, soapsuds, and sumac berries have all been used. Birds have also been found to place ants under their wings, and grackles are known to anoint their feathers with berry juice and ant with mothballs.

It is even more astounding when you imagine a bird individually cleaning every one of its feathers, which make up almost 10% of their total body weight. Even the Ruby-throated Hummingbird, a bird that has the fewest feathers of all, must clean approximately 940 feathers. Not that it will clean all 940 at each sitting, but it is a formidable responsibility, nonetheless. And how about the larger birds? A swan carries the highest number of feathers, approximately 25,000. While the number of songbird feathers ranges from 1,500 to 5,000. Meadowlark feathers number about 4,600, blackbirds possess about 4,900 feathers, and bobolinks carry about 3,200 feathers. The all-white Great and Snowy Egrets possess somewhere between 18,000 and 12,000 feathers, respectively. Considering the habitats they frequent, those are a lot of feathers to maintain in their pure white condition.

Our Lord's Candles Are in Full Bloom

APRIL 13, 1997

Almost everywhere you go in the countryside this spring, it is difficult to miss the tall, flowering stalks of our common yucca, the Spanish Dagger.

Some of these wonderful plants may reach 20 feet in height, although most are less than half that. The tall, white-flowering stalks are extremely attractive. The great masses of flowers have given the plant the name of "Our Lord's Candles," which is most appropriate. Other common names include "Spanish Bayonet" and "Palma Pita"; scientists know the plant as *Yucca treculeana*.

Our Spanish Dagger possesses thick, dark green or bluish green leaves that branch out in all directions from the stalk. Some of these fleshy leaves may be 40 inches long and 3 inches wide at the base and taper to a very fine, sharp point. They can easily puncture the skin or clothing.

The flowers, borne in dense masses, contain three outer sepals and three inner petals up to 2 inches long. After blooming in spring, often from February to mid-summer, leathery fruits appear. These may be 4 inches long and filled with tightly packed black seeds.

Coastal forms of Spanish Daggers tend to be taller with shorter leaves and a relatively slender trunk; more inland plants possess longer leaves with a shorter, stockier trunk.

Native Americans and Texas settlers utilized Spanish Daggers in numerous

Spanish daggers produce masses of flowers when blooming.

ways. Leaves provided coarse fiber that was made into sandals, baskets, rope, and cloth. They also were used for fencing and in thatching for walls and sides of huts. Young flower stalks, buds, and flowers were eaten raw, boiled, or pickled. The petals, often used in salads, are high in vitamin C. The fruits were also baked, peeled, stripped of fiber, and boiled down to a pulp that was then rolled out in sheets and dried; the material could then be stored and used like molasses on bread and tortillas. Prepared fruits also were fermented for a powerful beverage, and soap, known as "amole," was made from some yucca roots.

Our Spanish Dagger is but one of twenty yucca species known in Texas, according to Donovan Correll and Marshall Johnson's classic *Manual of the Vascular Plants of Texas.* Most botanists place yuccas in the lily family, while others lump them with agaves in the agave family.

The pollination of yuccas depends upon a tiny yucca moth that lays its eggs within the flowers, pollinating the plants by pushing pollen into the stigma tube in the process. In return, the moth larvae feed on the seeds.

But however they reproduce and wherever they belong in the botanical classification scheme, our Spanish Daggers are one of our most outstanding native plants.

Bats Keep Getting a Bad Rap

APRIL 16, 1995

Bats receive lots of negative attention in South Texas. Much of what has been written about bats, at least in the popular press, is not complimentary.

It seems these hairy creatures constantly find themselves in conflict with humans either by roosting in the wrong place or by getting blamed for all sorts of mischief, from getting caught in human hair to carrying rabies.

Although "cave" bats may often seek out a dark and secluded place (like an attic) to roost during the daytime, and bats do contract rabies (rarely) from other sick bats, they never fly into anyone's hair; that is an old wives' tale. Also false is the phrase "blind as a bat," as all bats possess good vision. All in all, bats are an important part of our natural environment and are valuable contributors to the ecosystem in a number of ways.

All of the bats that occur in our area are insect-eaters. They spend the better part of their nighttime hours flying about, capturing and eating flying insects of various kinds. A single Little Brown Bat can catch 600 mosquito-sized insects in an hour. How they do this is truly amazing, for bats actually locate objects with high-pitched sound waves, just like ships and planes use radar.

Their high-pitched squeaks—between 25,000 and 75,000 vibrations per second (humans can hear only about 20,000 vibrations per second)—bounce back from whatever they touch, so that bats can determine an object's direction, distance, velocity, shape, size, and even the texture of their prey—even objects as small as the width of a human hair. To test their echo-location ability, throw a tiny stone in the air near a flying bat, and, thinking it is an insect, it will dive after the falling rock.

Flying insects, tracked by echo-location, are captured either by the bat's wide-open mouth or by being swept up by its membranous, netlike wings and eaten in flight. They also may land on the ground to capture insects. Especially large prey may be taken back to the roost to eat at their leisure.

Approximately 70% of the world's bats (919 species) are insect-eaters that consume more than one-third of their body weight in flying insects nightly. The remainder (30%) are fruit- or nectar-eaters; a tiny percent of tropical species live on blood.

Of 32 bat species that have been found in Texas, only 7 of those have been found in the Golden Crescent, according to David Schmidley's comprehensive book, *Bats of Texas:* Eastern Pipistrelle, Eastern Red, Seminole, Northern Yellow, Hoary, Evening, and Brazilian (earlier known as Mexican) Free-tailed Bats.

All seven of our bat species are insect-eaters. Five species—Red, Seminole, Hoary, Northern Yellow, and Evening—roost primarily in trees and foliage, and the others roost primarily in caves and crevices. Three species—Pipistrelle, Evening, and Free-tailed—can often be attracted to artificial structures such as bat houses.

There is a growing interest in utilizing bat houses to attract these insect-eaters to where they are needed. Bat houses work very well if placed in suitable locations where the residents do not get overheated by our southern sunshine. Further information on bats can be obtained from Bat Conservation International, P.O. Box 162603, Austin, TX 78716.

Watch for Spittle from Spittlebugs

APRIL 18, 1999

Odd-looking masses of tiny bubbles are beginning to appear on various shrubs throughout South Texas. This strange stuff is spittle, a frothy material that provides a protective cover for insect nymphs that, in their adult stage, are known as Froghoppers, members of the insect order Homoptera. The adults are small hopping insects, rarely over ½ inch in length, that usually resemble tiny frogs. They normally are brown or gray in color, although some species have a characteristic color pattern.

The spittle is derived from fluids voided from the insect nymph or Spittlebug, the immature form of the Froghopper. Eggs are laid on plant stems during fall or early winter. When the eggs hatch in spring, the nymphs immediately disperse to a suitable site where they create spittle masses from the excess plant juices that they feed upon. Lashing its little tail, the nymph exudes a soapy liquid into which it pumps air by means of a curious bellowslike apparatus in its abdomen. As the bubbles are formed, the insect pulls and pushes them over itself in an orderly fashion until its soft body is completely covered. The nymph usually rests head downward on the plant, and as the spittle flows down and over its body it covers the insect, providing it with a moist, protective cover. The nymphs undergo several molts within the spittle, usually moving to another site and creating new spittle each time. Finally, after several molts taking several weeks, they emerge as winged adult Froghoppers that can no longer create spittle. The adults then go about their business of producing additional generations. Froghoppers have one to three generations each year, depending upon the species and latitude. Eggs of the last generation overwinter to hatch in spring.

It is relatively easy to find Spittlebugs in the nymphal stages. Simply find some spittle on plant stems, usually in lush, weedy places. Gently push the spittle aside—it is little more than air bubbles—and find the nymph inside, sitting aside the stem. There is usually only one Spittlebug per mass of spittle, but occasionally there may be as many as five. These immature Froghoppers are light green and about ⅛ inch in length. Adult Froghoppers can be found by bending down among lush grasses, brushing your hand across the grass, and waiting for the adult to jump on you. They are capable of making large hops and are brown, stout, oval insects about a ¼ inch long. Froghoppers look a good deal like Leafhoppers but are stubbier and lack the rows of spines that adult Leafhoppers possess on their legs.

Creating spittle is a distinctive characteristic of the nymphs of this group of insects. Besides providing a moist environment to keep the vulnerable nymph from drying out, the spittle is able to last for considerable time, even when exposed to heavy rains. It also provides camouflage and is distasteful to potential predators. Although the spittle itself is fairly easy to find, most predators are not willing to wade through the spittle to obtain such a tiny morsel.

Froghoppers feed on shrubs and herbs and normally have little impact on the plants. One eastern species feeds on clover, sometimes stunting the growth of the plants it feeds upon. Another species feeds on pines and can be a pest. Most of the Froghoppers and their nymphs that occur in our area of South Texas are little more than harmless, curious creatures, another insect in our fascinating world of bugs.

Bird Migration Is an Amazing Event

APRIL 20, 1997

Spring in South Texas is an exciting time of year! It is next to impossible to ignore the abundant migrating birds that are passing by. They include birds of every color and shape, many of which are old friends that haven't been seen since last spring. It only makes sense, then, that almost everyone I visit with this time of year seems to be filled with questions about bird migration. Here are answers to a few of the more common questions.

Where are the migrants going and where are they coming from? Most of our passing birds are neotropical migrants, species that spend their winters in the Tropics, from central Mexico to South America, and nest in North America, from Texas to Alaska. Some Arctic shorebirds that winter in southern South America and nest in northern Alaska travel a round-trip distance of well over 13,000 miles.

Most of the songbirds passing through South Texas are Trans-Gulf migrants that leave Mexico's Yucatán Peninsula in the early evening and arrive along the Texas Gulf Coast the following day (depending upon weather conditions), a distance of about 550 miles. Songbirds are able to fly nonstop for eighty to ninety hours. You can watch the night migrants passing overhead on a clear night by setting up a spotting scope aimed at the moon; the abundant birds appear as black specks.

Do all birds migrate at night? Most do, but many others, such as swallows and other insect-eaters that feed in flight, usually migrate during the daylight hours. We see many of these birds flying north over the fields and woodlands of South Texas in the month of April.

How fast do birds fly? Most long-distance migrants travel between 25 and 40 mph. Flight speeds vary with their activity. For instance, Purple Martins fly 27 mph, shorebirds fly between 45 and 55 mph, hummingbirds may fly up to 55 mph; and Peregrine Falcons can stoop at over 100 mph.

How high do birds fly? It varies with the topography, but 90% of all migrating birds fly below 5,000 feet above groundlevel. Many fly much lower so we are able to hear chips on a calm day or night. They tend to fly higher at night when flying overland.

Do birds migrate in mixed flocks? Mixed flocks of songbirds, ducks, and shorebirds are normal, but some others species, such as Nighthawks and Chimney

Swifts usually stick with their own species. In the fall, several raptor species can often be found within one area, but most hawks also stay with their own kind.

How do birds prepare themselves for migration? Most accumulate great quantities of fat as fuel for their long-distance flights. Many double their weight. The tiny Ruby-throated Hummingbird, weighing 4½ grams, uses 2 grams of fat to fly nonstop for twenty-six hours. A typical bird will lose almost 1% of its body weight per hour while migrating.

What is a bird's signal to migrate? Although the answer is complicated, a simple answer is the increasing hours of daylight in spring. You need not worry about your feeders preventing winter birds from leaving. Your bird food only helps the birds prepare for their journeys.

Gulf Coast Flyway Is the Center of Spring Bird Migration

APRIL 21, 1996

The last two weeks of April mark the peak of the spring bird migration, a subject that can not be overemphasized in South Texas. Millions of neotropical migrants are passing through South Texas, en route from their wintering grounds south of the border to nesting sites all across North America.

These neotropical migrants range from tiny hummingbirds, slightly larger warblers and vireos, mid-sized thrushes, to larger shorebirds and hawks, and even nocturnal species such as the Chuck-will's-widow and Whip-poor-will.

Those of us living in South Texas are better able to appreciate this migration than anywhere else in America. We live in the very center of the Gulf Coast flyway, the route used by the highest number of the northbound migrants. These birds enter South Texas either from along the coastal strip that extends south into Mexico or from across the Gulf of Mexico via the Yucatán Peninsula. While many of the daytime migrants, such as swallows that feed on the wing, migrate over land, others are nocturnal migrants that cross the Gulf.

Every evening during April and May, thousands of birds gather in great staging areas along the northern portion of Mexico's Yucatán Peninsula, waiting for the right signal to take off across the Gulf on one 550-mile-long flight to the Gulf Coast of Texas, Louisiana, and Mississippi. Leaving soon after dark, they arrive on our coast in mid- to late morning, often totally exhausted from their overwater flight. Depending upon weather conditions along the way and along the Gulf Coast, they may either continue on over the coastal strip to choice feeding grounds a few to dozens of miles inland, or, completed exhausted, they may literally drop out of the sky into the first part of the mainland encountered.

Places like High Island, north of Galveston, have a well-earned reputation for attracting birders in spring. When conditions are right, one can find hundreds of songbirds and other species among the oaks, often as many as 80 to more than 100 species. These tired Trans-Gulf migrants first rest a few minutes or longer, and then they begin to feed on the available insects, building up their body fat for the next leg of their journey.

But we need not go all the way to High Island to experience such fall-outs. They also occur within the Golden Crescent. Oak mottes from Aransas to Palacios with their abundance of spring caterpillars also provide recovery sites for the hungry migrants. Bennett Park, near La Salle, is one of the better nearby fall-out sites.

But one can appreciate the spring migration almost any place in South Texas. It begins to build in mid-March, increases through April, and slowly declines to mid-May. Even when birding only in my yard near Mission Valley, I can record approximately 30 bird species a day in March, 45 in early April, and 50 to 65 species later in the month. By visiting a dozen sites in Jackson, Calhoun, Victoria, and Goliad counties on april 29, 1995, local birder Mark Elwonger and two birding friends recorded 190 species in sixteen hours. Good proof that we live in a birding paradise.

Port Lavaca Tops List of Birding Sites

APRIL 23, 1995

Port Lavaca and surrounding Calhoun County was the centerpiece for the four-day (April 20–23) 1995 spring meeting of the Texas Ornithological Society (TOS).

Field trips were scheduled to sample all of the nearby habitats, ranging from a deep-water pelagic trip from Port O'Connor, to additional trips to

Matagorda Island, Indianola, Welder Flats and Seadrift in Calhoun County; Lake Texana and Bennett Park in Jackson County; Victoria's Riverside Park; and Aransas National Wildlife Refuge in Refugio County.

The TOS Convention, attended by more than two hundred participants from all across the state, was under the able coordination of Linda Valdez, president of the Golden Crescent Nature Club, and Doris Wyman of Port Lavaca, with help from lots of other club members.

TOS meetings are held only at choice birding sites in the state, and so it is only natural to assess the virtues of Calhoun County. In doing so, it is immediately obvious, at least in spring, that this area of the Texas Gulf Coast may be the single most productive birding area anywhere in Texas. And since Texas is the number one birding state in all of North America, that gives Port Lavaca and vicinity an exceptionally high rating.

As proof of this prestigious position, the highest count of bird species (211) was recorded in Calhoun County on the 1994 Annual North American Migration Count, a one-day bird count held each year on the second Saturday of May in counties all across North America. Nueces County was second with 170 species; Brazoria County was third with 157; Victoria County birders tallied 139 species.

Why is Calhoun County North America's numero uno spring birding site? First and foremost, the county is situated along the Texas Gulf Coast where it attracts two types of northbound migrants. Millions of the neotropical spring migrants follow the coast from Mexico through Texas and pass through Calhoun County; most do not branch off toward the west until they reach the major river corridors beyond.

In addition, Calhoun County is far enough north along the Gulf Coast that millions of Trans-Gulf migrants touch land there first. This is most obvious when thousands of songbirds suddenly descend out of the sky during the mid-morning hours in April and early May. After a 550-mile flight across the Gulf, the tired and hungry migrants will often permit easy viewing among the insect-filled riparian sites and oak mottes.

Furthermore, Calhoun County contains four rather distinct vegetation zones, each supporting its own variety of resident breeding birds. The Live Oaks, with nesting Red-bellied Woodpeckers, Blue Jays, American Crows, Carolina Chickadees, and Tufted Titmice, represent the Eastern Forests. Western influences, most evident by mesquite, add nesting Ladder-backed

Woodpeckers, Cactus Wrens, Curve-billed Thrashers, and Pyrrhuloxias. The South Texas Plains species include White-tailed Hawks, Buff-bellied Hummingbirds, Brown-crested Flycatchers, Long-billed Thrashers, and Olive Sparrows. And, finally, beach and wetland species add a huge variety of other birds to the mix, including a multitude of waders, shorebirds, gulls, terns, and marsh birds.

Is it any wonder that Calhoun County has attracted the TOS members to experience its amazing assortment of spring bird life?

Social Parasites of the Bird World

APRIL 24, 1994

Human beings are not the only animals that must put up with social parasites. Birds also must deal with fellow creatures that survive off the goodwill of others.

For birds, the principal culprits are the Brown-headed and Bronzed Cowbirds. Females of both cowbird species seek out other bird nests, usually those of smaller species such as warblers, vireos, and gnatcatchers, to deposit one to three of their own eggs. The host birds seldom recognize the cowbird eggs and hatchlings and raise them as their own.

The female cowbird is a remarkable creature with an extremely long reproductive period with very short intervals between egg-laying. Her laying cycle is adapted to take advantage of a continuous supply of host nests for about two months each spring. She has been called a "passerine chicken."

Female cowbirds may also carry off (and occasionally eat) one of the host's eggs when depositing her own egg. Although the cowbird egg is deposited about the same time as that of the host, the cowbird egg hatches in only about twelve days, sooner than those of the host. And since the cowbird nestling is larger and grows faster, it has a significant head start on the rightful nestlings. Sometimes it even shoves the smaller birds out of the nest in its fight for food being brought to the nest. The foster parents faithfully incubate all the eggs and care for all the nestlings, young cowbird included.

By the time the cowbird is ready to leave the nest, it may be larger even than the foster parents. Because the size difference is so great, an adult warbler or vireo may actually stand on the head of a cowbird youngster while feeding it. After leaving the nest, the young cowbird will continue to follow its foster parents about, begging incessantly to be fed. A week or two later, however, the young cowbird will leave to join flocks of its own kind.

Brown-headed Cowbirds reside in the Golden Crescent area year-round, while the slightly larger Bronzed (or Red-eyed) Cowbirds, that parasitize even larger species, normally migrate south into deep South Texas or Mexico for the winter. Brown-heads spend their nonbreeding season in flocks that sometimes number in the thousands, often accompanied by European Starlings and Red-winged Blackbirds. They feed on grain in open fields and at feed lots.

About the same size as a common Red-winged Blackbird, male Brown-headed Cowbirds are black with a brown head; females are a dull brown. The wheeling flocks of cowbirds often make loud screeching noises, but individuals possess a slurred squeaking and gurgling call, like pouring water.

Brown-headed Cowbirds have been resident within the Golden Crescent area for years, but they once were found only on the prairies with the vast herds of American bison (buffalo). The Plains Indians called them "tick-birds" because of their habit of eating ticks off the bison. Their principal foods, however, were the seeds found in bison droppings. But with the great decline of the bison, cowbirds were almost obliterated from the American scene. A resurgence of the species occurred with the arrival of cattle and the increase in cattle herds throughout the West. Along with the conversion of forests to farms and pastures, the Brown-headed Cowbird revival has continued to this day.

There are probably more cowbirds now than any time in the past, but the increasing numbers are having dramatic, negative effects on several songbirds all across the country. In Texas, serious declines of the Golden-cheeked Warbler and Black-capped Vireo in the Hill Country, partly attributed to nest parasitism from Brown-headed Cowbirds, have led to the listing of these two songbirds as endangered.

Brown-headed Cowbirds have been recorded as successfully parasitizing more than 200 species of birds, ranging in size from the tiny Ruby-throated Hummingbird to the larger Blue-winged Teal. They represent one of the most successful comebacks ever in the bird world, but their overabundance in some areas has created other problems.

Spring Moths Are Drawing Attention

APRIL 27, 1997

Numerous types of moths fly in the month of April, and at least two of those—the common Sphinx and the huge Polyphemus Moths—cannot help but attract one's attention.

Sphinx Moths, sometimes called "hummingbird moths" or "hawk moths," are active during the daylight hours, especially in the mornings and evenings. They usually are seen feeding on flower nectar in our gardens or along the roadsides. Their behavior of hovering over flowers, with their long proboscis, like the beak of a hummingbird, probing the flower heads has given them the name of "hummingbird moth." And, because of their similar behavior, these moths are sometimes even misidentified as hummingbirds. Their "Sphinx" name relates to the sphinxlike appearance of the larvae, known as "hornworms," due to the conspicuous or spikelike process on top of their eighth abdominal segment.

Sphinx Moths possess heavy bodies, sometimes spindle shaped, with long, narrow front wings. They possess a very rapid wing beat, again similar to hummingbirds. Most species are brown and black with streaks of rose or reddish colors.

Polyphemus Moths are huge—ten to twelve times larger than most Sphinx Moths. And they are far more difficult to find, since they normally are active only at night. Males, however, are sometimes active during the daytime when searching for a mate. Otherwise, they are most often disturbed from the bark of a tree or among vegetation where they rest during the daylight hours. Then they flap slowly away, almost batlike, out of danger.

Polyphemus Moths can hardly be misidentified because of their huge size (wingspan of about 5 inches) and key field marks: soft buff or tan with pinkish shades, and with one obvious eyespot, usually bluish with a yellowish center, in the center of each hindwing. They also have large, noticeable, featherlike antennae.

Their name comes from the one-eyed giant Polyphemus of Greek mythology. He was the Cyclops that Odysseus blinded. To scientists, they are known as *Telea polyphemus*, a member of the Saturniidae family of giant silkworm moths. The larvae feed on leaves of oak, hickory, elm, maple, and birch trees. They are huge, fat, accordion-shaped grubs that are armed with conspicuous tubercles or spines. And they also possess the very strange habit, when disturbed, of raising up and clicking their jaws. Although they may appear dangerous, they are not.

These two moths are just part of our fascinating natural world in South Texas.

Sabal Palm Is Native to South Texas

APRIL 28, 1996

South Texas is home to one of America's most attractive trees, the native Sabal Palm. Yet, too many people either take it for granted and ignore it or have little or no knowledge about it.

The species—*Sabal mexicana*—was once thought to occur in Texas only in the lower Rio Grande Valley, but it actually is native northward along the coast into Jackson County and inland to Austin. And it also has been widely cultivated and planted in cities and towns throughout its range.

There are only two native palms in South Texas, and the Sabal Palm is our only tree-sized native palm; it can reach to 50 feet in height. The much smaller Bush or Dwarf Palmetto (*Sabal minor*) is rarely more a few feet tall, although some may reach 15 feet. Young Sabal Palms are sometimes difficult to tell apart from palmettos, but young Sabals possess longer leaves and stems, the leaves are highly filiferous (stringy), and they are lighter in color. Palmetto leaves are seldom stringy in character. And adult Sabals possess green fan-shaped leaves that may be 5 to 7 feet long and equally as wide.

Sabal Palms were reported for the Rio Grande Valley by the very first explorers, and the French explorer Sieur de La Salle recorded their presence on Garcitas Creek as early as 1685. During Texas's early history, Sabal Palms were commonly used in wharf construction because they were immune to shipworms. Hundreds were utilized for piles for the wharves at the port of Indianola, and later they were widely dug up and sold for ornamentals. Many of the attractive Sabal Palms throughout the older portion of Victoria were planted there during the 1930s and 1940s.

It seems that we no longer appreciate our native plant life as we once did. It may be the fault of the nursery industry that can profit most from selling the more exotic and fast-growing species.

Today Sabal Palms continue to be destroyed throughout their range by developers or by unknowing individuals. Too often our native species are removed because they seem to be in the way, and many times other non-natives are introduced in their places. The nonnatives require much more care, or the more hardy introductions may spread into adjoining areas and eventually overcome some of our more desirable natural species.

The native Sabal Palm is one species that we would probably want to introduce if it were not already native. We should take advantage of its presence, encourage its reintroduction, and protect areas where it currently exists.

Swallows Are Passing through the Countryside

APRIL 30, 1995

Throughout the month of April, 5 species of swallows—Tree, Northern Rough-winged, Bank, Cliff, and Barn—pass through South Texas en route to their nesting grounds to the north.

At the end of April, except for an occasional late flock of migrants, we are left only with the two species that nest in our area, Cliff and Barn Swallows. Of course, Purple Martins are also swallows and they are still with us, and Cave Swallows also occur at a few sites in Goliad County. But, in general, only Cliff and Barn Swallows are commonplace.

Both of these swallows utilize human-made structures to build their mud-pellet nests. Cliff Swallows are most numerous and are especially common at concrete overpasses, culverts, and similar structures. They also utilize natural cliffs when available. These swallows, sometimes known as "square-tailed swallows," in comparison with the long, forked tails of Barn Swallows, usually occur in huge colonies, occasionally up to 1,000 pairs. Each pair constructs a nest composed of hundreds of tiny mud pellets, gathered and rounded in their bills and plastered together as building blocks in a retort-shaped structure with an extended opening at the lower end. The entire nest, including the interior that is lined with grass and feathers, is built in five to fourteen days. They sometimes simply rebuild or repair a previous nest.

The long-tailed Barn Swallows also build mud-pellet nests but seldom are part of a large colony. And the Barn Swallow nest, very different from that of the Cliff Swallow, is open at the top. They, too, utilize human-made structures and may even nest on the same bridge or cliff being used by Cliff Swallows. Barn Swallows are most distinguished by their long, deeply forked tail, but they also sport a reddish to buff throat and chest and blackish back. Adult Cliff Swallows, on the other hand, possess a buff rump and forehead, cinnamon cheeks, blackish throat, grayish underparts, a black cap, and a square tail.

Both of these swallows are neotropical migrants that spend their winter months in the Tropics. Cliff Swallows winter as far south as South America, from Paraguay south to Brazil and central Argentina. Barn Swallows normally winter in Mexico and south to Central America. But both are some of the earliest breeding birds to return to their nesting grounds in South Texas; Barn Swallows arrive in February, and Cliff Swallows appear the first of March. They will remain in South Texas throughout the summer months and may even nest twice when conditions are right.

Both species feed on the abundant insects that they capture in flight with their large open mouths. Their dexterity is a thing to behold.

May

Chimney Swifts Are Back from the Amazon Basin

MAY 3, 1998

Another of Mother Nature's amazing creatures has returned to its ancestral nesting grounds from its wintering grounds in South America. Chimney Swifts are again flying about South Texas neighborhoods, consuming tons of insects and thrilling us with their amazing aerial gymnastics. Especially during courtship, they fly in twos and threes, circling and diving with utter abandonment. And then they soar with their wings held in a V-shape pattern, a behavior that occurs most often with their mates.

Chimney Swifts have often been described as "flying cigars," due to their streamlined body shape and torpedo-like flight. That flight is quite different from other birds, consisting of quick flickering wing beats and then sailing with wings held out motionless. At first glance, Chimney Swifts might be misidentified as swallows, but swifts are not swallows at all. They are members of the same order of birds as hummingbirds, Apodiformes, a Greek word meaning "without feet," which is a misnomer because they do possess feet. They are delicate yet strong enough in flight to break small twigs off trees for nesting material. But like swallows and bats, their food and water is obtained in flight. Their diet consists primarily of flying insects and spiders on silken threads, and they skim water surfaces to drink.

Another atypical habit for birds is the Chimney Swift's use of chimneys for nesting. They once nested only in dark tree trunks and similar cavities. Now, they build their half-saucerlike nests on the inside of chimneys, using their gelatinous saliva to cement the tiny twigs together and to the wall.

Chimney Swifts have learned to utilize chimneys for nesting.

John Tveten, in *The Birds of Texas*, describes a nest he examined as a youngster: "The nest contained 130 twigs, each of them one and one-half to two inches long and the thickness of a toothpick or matchstick. All were laid parallel along the longer axis of the nest, forming a half-saucer about four inches across and two inches deep. That bowed shelf had been glued along one edge to the inside wall of the chimney."

A single chimney will be utilized by only one nesting pair, although other individuals that are helpers only—usually last year's young—may also be present. A clutch consists of three to five youngsters. Until they are twenty-eight to thirty days old, they are unable to feed themselves, but soon afterward they accompany their parents on their feeding excursions. Then, several family groups may be seen feeding together. Sometimes those flying flocks may number in the hundreds.

Although these "flying cigars" are numerous in Texas and usually closely associated with humans, they are among our most misunderstood birds. Perhaps that is largely due to their use of the dark interiors of chimneys and other chimneylike structures for roosting and nesting. Perhaps folks that fear bats have the same fear of other species that retire to the "dark side." But that is silly because Chimney Swifts are not only one of our most fascinating birds, but also one of the most beneficial species.

In recent years there has been a rather significant decline in Chimney Swift populations throughout their range, which extends throughout Texas; they are less common in far West Texas and in the Lower Rio Grande Valley. Because of their decline, Paul and Georgean Kyle, directors of the Driftwood Wildlife Association, have established the North American Chimney Swift Nest Site Research Project in cooperation with the Texas Parks and Wildlife Department. Its purpose is to develop a better understanding of the species and to enhance their numbers. The project has even developed Chimney Swift towers (8–20 feet deep and with an internal diameter of 11–20 inches) that can be constructed of wood to provide nesting sites for these misunderstood birds. Such sites are utilized by nesting as well as migrating swifts. For additional details about the project, towers, and Chimney Swifts in general, you may contact the Kyles at Driftwood Wildlife Association, P.O. Box 39, Driftwood, TX 78619; 512-266-2397; DWA@concentric.net.

Greening Mesquites Are Spring's Most Reliable Indicators

MAY 4, 1997

In talking to dozens of friends and colleagues in recent weeks about what is considered the most reliable indicator of spring in Texas, the only thing that everyone seemed to agree upon was the greening of mesquites.

All kinds of harbingers of the new season were mentioned. They ranged from returning Purple Martins and Scissor-tailed Flycatchers along the central Gulf Coast, Yellow-throated Warblers in the Pineywoods, and Black-capped Vireos and Golden-cheeked Warblers in the Hill Country. Further north, spring heralds include flowering Trout Lilies and Elbowbushes. In the Panhandle, the songs of Western Meadowlarks signify the arrival of spring.

When discussing this topic with Lorie Black of Abilene, she told me about a most pertinent poem that appeared in the *Abilene Reporter-News* on March 21, 1996. Then she sent me a copy. Apparently, each year since 1939, this poem, written by Frank Grimes (editor of the *Abilene Reporter-News* from 1919 to 1961) is republished. It is worth repeating here.

Old Mesquites Ain't Out

We see some signs of returning spring,
The redbird's back and the fie'larks sing,
The ground's plowed up and the creeks run clear.
The onions sprout and the rosebud's near;
And yet they's a point worth thinkin' about—
We note that the old mesquites ain't out!

The fancier trees are in full bloom—
The grass is green and the willows bloom,
The colts kick up and the calves bend down;
And spring's a-pear-ently come to town;
And yet they's a point worth thinking about—
We note that the old mesquites ain't out!

Well, it may be spring for all we know—
There ain't no ice and there ain't no snow.

It looks like spring and it smells so, too.
The calendar says it plenty true—
And still they's a point worth thinkin' about—
We note that the old mesquites ain't out!

And so it is! In spite of our returning Scissor-tails and all the fresh green growth, including flowering Texas Bluebonnets and Huisache, spring in Texas really never happens until the mesquites produce green leaves.

Larvae Pits Are Ants' Downfall

MAY 7, 1995

> *"Little fish have bigger fish that feed on them and bite 'em,*
> *and big fish have still bigger fish, and so, ad infinitum."*

When Jonathan Swift, best known as the author of *Gulliver's Travels*, penned the above lines, he probably was thinking only of fishing, and not about nature's pyramid of life. Yet, Swift's words very well relate to one of nature's most dramatic encounters, an Antlion's capture of its prey.

Dozens of round pits, each about one inch in depth, can often be found in the fine dirt at the edge of cliffs and buildings at this time of year. Each pit, an inverted pyramid, is rather steep sided, so that when an ant walks across the pit, it will slide backwards into it and be forced to climb out. This action will send loose dirt tumbling to the bottom. Suddenly, the dirt at the bottom of the pit will be thrown upward, slowing the ant's progress so it will again slide backward into the bottom of the pit. It is then that the ant is suddenly grabbed and eaten by one of nature's most specialized predators, the Antlion, sometimes called "Doodlebug."

Recently, I watched this activity for several minutes, as an ant fought furiously to escape by ascending the loose dirt of the pit. But just as it was about to reach the top, the tiny creature hiding underground below the center of the pit flipped dirt upward, and the ant again slid backward. Then, just as the ant was starting its uphill climb again, two tiny pinchers grabbed the ant from underneath, and I watched as the ant was slowly dragged out of sight. I had observed an Antlion capture its prey, and I knew that below the surface, the predatory Antlion was feeding on the unfortunate ant.

Such acts of nature occur thousands of times a day, all across the southern states. It is simply the method that this particular kind of insect has developed to capture prey. And these kinds of actions are readily available to anyone with the curiosity to watch.

Antlions are tiny but fierce creatures that are surprisingly common around our homes. Actually the larval form of a species of delicate, four-winged insects (family Myrmeleontidae), the Antlion looks very much like a Lacewing or Damselfly. The adult possesses two pairs of long, narrow, multiveined wings and a long, slender abdomen. They are rather feeble fliers and often are attracted to light after dark.

Eggs, laid in the dirt, produce a small colony of soft-bodied larvae with strong sicklelike jaws. Each larva digs a funnel in the dirt to aid in capturing ants. After feeding on its prey, the larva throws the remains clear of the pit, and the pit is repaired for the next meal. After some growth, the larva builds a rough cocoon of sand and silk, from which it will soon emerge as a winged insect.

Anyone interested can find his or her own Antlion pits and watch nature in the raw.

Butterfly Development Is Nothing Short of a Miracle

MAY 10, 1998

No other group of animals undergoes such a remarkable transformation as do butterflies. The butterfly's metamorphosis from egg to larva, caterpillar to pupa, or chrysalis to adult butterfly is truly remarkable. Butterfly life history is one of nature's most amazing happenings. And as spring and the summer months descend upon us, that miracle is all around, for all of us to see and appreciate.

Butterflies have a life cycle that is called *complete metamorphosis* because it includes four stages. Butterfly eggs are tiny things that can be round, spherical, or bun-shaped and may come singly, in small clutches, or in huge masses of up to 500 eggs that are attached by a gluey substance. The eggs are laid on a plant that the hatched caterpillar can utilize as a food source.

The eggs can be laid on top of or beneath a leaf, on a twig, or even at the base of grass, depending on the butterfly species.

Hatching can take anywhere from a few days to a full year, again depending upon the species. For instance, Falcate Orangetip butterflies fly only from March into May, during which time they lay eggs and live as larvae; the remainder of the year they occur only as chrysalides. However, most butterflies we see during the year pass through the four stages in only a few weeks, and so we see fresh specimens constantly during the warmer days of the year.

Butterfly larvae, or caterpillars, are true eating machines that spend the majority of their existence consuming plant materials; the exception is the Harvester Butterfly larva that feeds on aphids. The body of a caterpillar is divided into three parts: head with a pair of simple eyes, mouth, and large jaw (mandible); thorax, with three segments containing three pairs of true legs for moving about; and abdomen, with ten segments containing five pairs of prolegs, built like suckers to aid in clinging to various materials. The jaws not only can tear plants apart but also assist in transporting food to the mouth. This eating machine's entire purpose is to convert plant or animal tissue into butterfly tissue.

Like all arthropods, the butterfly caterpillar grows by shedding its skin periodically, whenever the new exoskeleton develops and hardens underneath. Once the new exoskeleton is formed, the caterpillar breathes in extra air and splits the old outer skin down the middle, and simply crawls out of its old skin. This process is called *molting*.

Finally, when the caterpillar reaches its maximum size, it finds a safe location and spins a form of silken mat, often with a silken thread or girdle as a safety belt. It then hangs upside down and spins a silken sheet, not a cocoon (only moths spin a cocoon), on a leaf or other object. This time, when shedding its old skin, it changes into a chrysalis, an immobile stage in which it undergoes a massive reorganization. This transformation takes a week to several months, depending on the species and the time of year, and includes both internal and external organs.

Its emergence as an adult butterfly is one of Mother Nature's most incredible feats, going from caterpillar to butterfly, complete with small wings and an oversized body. On emergence, it quickly pumps fluid into the wings from the body that then shrinks to its normal size. The adult has also developed a proboscis (a long coiled suctioning tube for feeding) and six true legs. Emergence usually occurs in the early morning when humidity is high, temperatures are relatively low, and predators are less active. The first flight usually occurs in the afternoon.

Adult butterflies must have food and water to continue their life process, so they feed on nectar and pollen from various flowers and often also obtain water and nutrients from various sources such as rotting fruit, carrion, dung, and wet soil.

Butterfly courtship involves a rather complex behavior, involving recognition of the opposite sex by wing pattern and pheromones. Pairs often go sailing high in the air in courtship flights. Mating can continue for several hours, but then the female must find the suitable plant species on which to lay her eggs. Once she lays her eggs, the process of transformation from egg to adult butterfly begins over again.

Ants Are Highly Social Creatures

MAY 11, 1997

The approximately 9,500 species of ants, a population that has been estimated at a million billion individuals, make up more than a quarter of the world's total biomass. Although ants and the other social insects—including termites, wasps, and bees—make up only 2% of all known insects, ants alone weigh four times as much as all the birds, amphibians, reptiles, and mammals combined.

According to sociobiologist Edward O. Wilson, the reason for this superiority is their amazing social organization. In a fascinating little book, *In Search of Nature,* Wilson explains that an ant colony is a "kind of superorganism—a gigantic, amoeba-like entity that blankets the foraging field, collecting food and launching forays to engage enemies before they can approach the nest. At the same time they care for the queen and the immature pupae. . . . They accomplish all of these things with a high efficiency by a division of labor. Most importantly, they do them simultaneously."

Wilson provides a number of examples of their organization. For instance, in the Southwest deserts when the nest of a rival is discovered, fellow colony members surround the nest and dump pieces of gravel into it, closing off the exit and eventually burying the rivals beneath the rubble. The workers of one ant species in the Malaysian rain forest possess grotesque glands loaded with a sticky toxic chemical at the base of the mandibles. When confronting an enemy, these workers are able to contract their abdominal muscles and explode, like walking grenades. One of these ants can trade its life for the lives of several enemies.

Another reason for ants' success is their ability to "maintain the nest as a climate-controlled factory within a fortress." Some female workers are responsible for maintaining the optimum humidity. Such a worker will climb a shrub to gather morning dew that she then allows workers to drink, or she daubs part of her burden onto a cocoon or passes the rest to thirsty larvae. During dry periods, she places water drops "directly on the ground inside the brood chambers, keeping the soil and air moist." Some ant species store water, nectar, and dissolved fats inside their abdomens, which will expand like balloons. When returning home, they pass liquids to other colony members by mouth-to-mouth regurgitation.

Ants also control foraging space around the nests, and they are "able to bequeath the nest, which is very expensive to produce in terms of energy, together with the territory, to later generations."

Wilson points out a few other differences between human and ant societies. Humans orient ourselves and communicate chiefly by sight and sound, while ants do so largely through taste and smell. And most interesting, while humans send their young men to war, ants send their old women.

Yellow-flowering Retama Is a Symbol of Early Summer

MAY 12, 1996

The fragrant and bright yellow flowers of our native Retama are starting to appear. There is no better indicator of summer. This lovely little tree, with thorny, green bark, can grow to 35 feet high. Its drooping foliage, rounded crown, and compound leaves (tiny leaflets on a long flat stem) also help with identification. And by late summer, light brown to reddish, narrow pods, 2 to 4 inches long, appear.

Our Retama, known to scientists as *Parkinsonia aculeata*, occurs in Texas from the Gulf Coast west to El Paso, mostly in sandy areas. A closely related tree known as Texas Paloverde, or *Cercidium texanum*, occurs from Del Rio west along the Rio Grande to El Paso and westward through Arizona. Texas Paloverde has smaller pods and a shorter flower stem. The two flowering trees look very much alike. In Mexico they collectively are known as "Palo Verde" (Spanish for "green stick"), due to their green bark. Also in Mexico, the two trees are often called "lluvia de oro," meaning "shower of gold."

Both trees are members of the pea or legume family that comprise over 500 genera and more than 10,000 species in all parts of the world. Retamas prefer moist sites and often are common in riverbeds and at springs. In the Texas Big Bend country, the presence of a Retama is often a reliable indicator of water.

Because of their attractive appearance, Retamas often are used for landscaping. Ornamentals rarely grow more than 20 feet in height. Also due to its all around attractiveness, Retama has been named the city tree of Corpus Christi. Flowering plants also make good bee-trees, due to their sweet, nectar-laden flowers. And hummingbirds are attracted as well.

Paul Cox and Patty Leslie in their book, *Texas Trees: a Friendly Guide*, point out that "livestock browse the foliage and branches during hard times. Bees are attracted to the flowers, and the pods are sought after as food by deer and other animals. In earlier times the pods were pounded and made into a course flour by Indians. In Mexico a tea brewed from the branches and leaves is used in the treatment of diabetes and as a fever remedy."

Whether the Retama is a wild tree growing along the riverbeds or is maintained in our gardens, it is among our loveliest and most cherished plants.

Barred Owl Youngsters Are Out and About

MAY 14, 1995

Barred Owls already are escorting their youngsters around their territories. After being rather quiet for the last several weeks, they are again calling to one another in their wonderfully deep hooting notes.

Barred Owl calls are marvelous! They usually occur in a series of eight accented hoots, ending in "oo-aw," with a downward pitch at the end. The eight notes have given them the name of "eight-hooter" in some areas.

Sometimes their calls phonetically can sound like "who cooks for you? who cooks for you-all?" Other descriptions include "howWHO-ha-WHOO! . . . howWHO-haWHOOAaahh!" But whatever it may sound like, the Barred Owl's deep hooting calls are one of nature's most memorable sounds. And when young are present, they may communicate with a wide range of sounds that include wails and moans to cackles, hisses, and laughs.

This mid-sized owl, with a range that extends from British Columbia east to Quebec and south to the Gulf Coast, prefers riverine habitats. The abundant floodplains that persist within the Coastal Bend offer plenty of choice sites.

Barred Owls produce an amazing assortment of calls.

Although it is most often detected by its wondrous songs, usually after dark, the Barred Owl also will call during the morning hours and on dark, overcast days. It occasionally is seen out and about during the daylight hours, especially in May when the adults are actively hunting to feed their two or three youngsters. Sometimes an adult will be followed by one of its fluffy babies.

Barred Owls prey on a wide assortment of creatures, from rodents and other small mammals, to frogs and toads, snakes and lizards, other birds, and a variety of invertebrates. They normally hunt at night, using special hearing. Their ear openings are offset to help in locating prey by triangulation. They are able to locate even faint sounds with amazing accuracy.

Owls also have specially adapted wing feathers that are serrated rather than smooth; this adaptation disrupts the flow of air over the wing in flight, eliminating the vortex noise created by airflow over a smooth surface. This allows the predator to approach its prey with little if any sound. They also can see reasonably well at night, even on the darkest nights. Owl eyes are dominated by rods, rather than cones, that are receptors and able to function in very dim light. Barred Owl eyes, therefore, are completely dark.

Barred Owls are normally 16 to 24 inches tall with a wingspan of 38 to

50 inches. They are somewhat smaller than the more robust Great Horned Owls that prefer open ranch land. Also, Barred Owls lack the ear tufts of the Great Horned and possess a large rounded head. Their name comes from the barring on their throat and upper breast, although they possess a streaked chest and belly. They normally are a rich brown color, although pale individuals also occur on occasions.

But for all of its appeal, the Barred Owl's characteristic hooting calls are its most enduring feature. Anyone who has not experienced its rhythmic, emphatically delivered "howWHO-haWHOO! . . . howWHO-ha-WHOOAaahh!" songs after dark along the Guadalupe or San Antonio rivers, or at numerous other sites in South Texas, has missed one of life's most memorable events.

A Family Disaster

MAY 17, 1998

It occurred at midday and was totally unexpected. For more than a month I had been observing whatever activities I could in the time available. The previous year had been the first in eight years since I had initially offered a home for my guest that she had taken advantage of my hospitality.

Last year she produced a single offspring, and Betty and I had taken photographs of that youngster as if it were our own. So this year when she again moved in, I was excited and was anticipating a similar happy ending, but our experience was very different.

The incident began when a Fox Squirrel approached the nest box which our guest, an Eastern Screech-Owl, was occupying. My observations of what occurred were accidental and inconclusive. As the squirrel touched the wood nest box, the Screech-Owl, apparently detecting the squirrel's presence, suddenly exploded from the box and flew into an adjacent Live Oak tree. I had no idea if the squirrel had bothered the nest site earlier. I then watched as the squirrel entered the box, where it remained for about twenty seconds. On exiting, it scampered down the tree, and the Screech-Owl immediately flew back to the box and entered. The owl was out of sight for only a few seconds before it reappeared at the entrance, where it remained, looking about in all directions. It appeared extremely nervous.

Within about ten minutes the squirrel returned, climbed the tree and approached the nest box again. This time, instead of flying off when it had the opportunity, the owl dropped back into the box out of sight. The squirrel immediately entered and remained inside out of sight for about fifteen seconds. As the squirrel left the box, several feathers flew out, and through my binoculars, I could see some feathers clinging to the squirrel's mouth. I did not, however, detect any blood. The squirrel immediately ran down the tree and scampered away into the dense brush away from the house.

I was extremely upset! My "own" Screech-Owl had apparently suffered foul play from another of my native yard creatures. I hoped against hope that the owl had adequately defended itself and that the squirrel had left the scene in retreat rather than victoriously. I continued to watch the nest box for almost an hour but observed nothing more that might suggest that the Screech-Owl had survived the encounter. And for several days afterward, I watched for it during the morning and evening when I had normally seen it perched at the exit hole. Nothing!

I finally checked out the nest box a few days later. I discovered one or two recently broken eggs but no dead Screech-Owl. Ants were still present, feeding on the egg remains. What about the Screech-Owl? I'm not sure. It may have survived the confrontation. I suspect it was injured but managed to leave the box after dark. It clearly was no longer utilizing the nest box.

Whatever had actually happened, both my wife and I were upset. But yet, I had watched an event that undoubtedly occurs in nature all the time. We humans look upon such events with our anthropomorphic perspective and rarely think beyond that scope. Such a predator-prey relationship, which is part of the great scheme of life, is all around us, but it hurts when it applies to some creature to which we relate.

Twig-girdler Beetles Have Been Hard at Work

MAY 23, 1999

Hundreds of twigs from my Cedar Elm tree have littered the ground in recent weeks. This clutter has been the work of Twig-girdler Beetles, a species of Longhorned Beetle (family Cerambycidae). Although a Twig-girdler

Beetle is only about 1 inch long with antennae about the same length as its body, it is able to cut twigs that may be as much as ½ inch across. Evidence of their work is fairly easy to detect; the cut twigs are all neatly severed with a blunt edge, rather than being jagged, as if the twig were torn off by the wind or by a bird. Also, the newer twigs still support leaves, although these will wither and die soon afterward.

Twig-girdler Beetles spend their winter as larvae within fallen twigs. In spring, the larvae pupate within the stems, and the adults emerge a few weeks later. After mating, female Twig-girders fly into a nearby tree, lay their eggs on the tips of twigs, and then crawl down the twig and cut a circle around it so only the tip of the twig remains. The adults then fly off. The larvae develop within the twig by eating the wood over a period of two or three years, during which time the twig usually falls off the tree. Oftentimes the twigs will be blown off the tree soon after the twigs are girdled, which is what we are finding at present. Twig-cutting apparently creates better conditions for certain stages of the insect's development. The adult Twig-girdler kills the twig in which the eggs develop, presumably creating conditions that make the wood more palatable to the larvae.

Longhorned Beetles are a huge group of more than 24,000 species worldwide, with 1,100 in North America. They range in size from ⅛-inch species to *Titanus giganteus* of Brazil, a huge, reddish brown creature that may be 5 inches long and 1½ inches wide. Some species possess antennae that may be two to four times the length of the body. One 3-inch species that lives in New Guinea possesses antennae that may grow to 7 inches in length.

All of the Longhorned Beetle larvae feed on wood from either live or dead trees. During this larval, or grub, stage they are sometimes considered delicacies by various native peoples. They are especially prized in Australia and South America. The larvae are extracted and toasted until brown and crisp, somewhat like certain cocktail snacks.

Since Longhorned Beetles usually fly at night, they are not regularly encountered. Finding these fascinating creatures in their natural settings will require careful observations, since most are about the same coloration as the woody materials that they inhabit. And to experience a tasty larva will require even more effort and considerable patience to examine a cut twig, extract one of the tiny inhabitants, and toast it to a crispy brown. Remember, it's a special treat somewhere!

Fireflies, Nature's Evening Sparklers, Light up the Nights

MAY 21, 1998

Every spring I am amazed all over again when the yard is suddenly filled with glowing lights from fireflies! I can't help but get excited for the umpteenth time about this marvelous work of nature. What child has not shared that amazement? Who has not captured one of these fascinating insects to watch it glow up close? Who has not wondered about these tiny creatures that turn our yards into miniature firefights?

The appearance of fireflies, sometimes known as "lightning bugs," are a sure sign that the winter months are behind us and summer is not far off. Fireflies appear only when the evening temperatures allow these cold-blooded insects, actually a soft-winged beetle of the Lampyridae family, to become active. The majority of the 2,000-plus members of the Lampyridae family (there are about 60 kinds in North America) possess luminous organs on their abdomens. They all are able to glow at will, usually starting at dusk and continuing until about midnight, generally as part of courtship to attract a mate.

Our fireflies are most numerous in humid yards and similar locations. The male firefly beetle is about ½ to ¾ inch in length, with a pair of fairly long antennae and a yellowish abdomen. Females look more like larvae than beetles. Although both sexes glow, males are capable of shining about twice as bright as the females.

Each of the more than 2,000 forms of fireflies possesses a slightly different glow pattern or code of signals. The specific code is not so much the light (either color, brilliancy, or length) as it is the length of the intervals. Each species glows for a specific length of time, with a specific interval between glows. The males of some tropical fireflies actually flash in unison, and the tropical nights can actually brighten as if a great light is being turned on and off. You can experiment with a penlight by copying the glow intervals to solicit a response.

The light itself is derived from fatty tissue in the beetle's abdomen called "luciferin." The light is produced when the insect takes in air through tiny ducts. When the air reaches the luciferin, it is instantly oxidized, releasing the energy as cold light.

During the daytime, the lampyrid beetles and their larvae are usually found on vegetation where they prey on smaller insects and larvae, small arthropods, and snails that they find in old leaves, humus, or on the ground. But at dusk, the males begin their nocturnal signals, flashing their unique lights in such a way that their mates are soon flashing back. And the night becomes magical!

Mesquite Trees Have Numerous Uses

MAY 26, 1996

Already, mesquite trees are covered with bright green leaves, and some individuals also sport green beans. Those long, thin beans will fatten up in a few weeks, and then they will be harvested by a variety of wildlife. Mesquites have long provided early desert peoples and a myriad of wild animals with valuable shelter, warmth, and food. Few native trees grown on arid land are as valuable as the abundant Honey Mesquite!

Scientists know the mesquite as Honey Mesquite (*Prosopis glandulosa*). They chart its range throughout Texas (except the Pineywoods) and from eastern New Mexico, Oklahoma, and Kansas, south into Mexico throughout the states of Coahuila, Chihuahua, Sonora, and west to Baja California. They also claim that it is more numerous today than anytime in the past, having increased tremendously in abundance over the past 130 years, probably as a result of disturbed rangelands. It is easily disseminated by seed, especially after passing through the digestive tracts of cattle.

Mesquite also has the ability to survive in extremely arid environments, so long as their taproots are able to reach moisture, sometimes as deep as 50 to 60 feet. The plants also are extremely hardy and not easily damaged by insects and diseases. And when cut to the stump, they usually are able to survive and continue to grow.

Native Americans utilized the entire bean for meal, pounding it into a fine powder; only the woody husks of the seeds were rejected. The powder was then sprinkled with water and packed in close layers into a pot, buried in the sand exposed to a hot sun, where it baked for several hours until the mass acquired the consistency of a soft brick. The "fried bread" contained a gummy sugar which dissolved in the mouth, and it was extremely nutritious. In addition, an intoxicating beverage was made from fermenting the meal. A nonintoxicating beverage was also made from the strained cooked beans.

Gum that exudes from the bark was used to mend pots or was eaten as a candy. New green beans were chewed as a sweet. Mesquite bark, when rubbed, pounded, and pulled until it became soft, was once used as diapers for babies and skirts for women. The leaves and beans are eaten by many wildlife species. Mesquite wood has historically been used for fuel, fence posts, railroad ties, and wagon wheels, hubs, and spokes. More recently, the wood has been used for many building purposes, such as attractive flooring, furniture, and gunstocks. Mesquite trees also are popular for landscaping, and now the wood has become popular for barbecuing.

What's more, an organization—Los Amigos del Mesquite—has developed to promote the tree's many useful aspects. I know of no other tree that has so many dedicated admirers.

May Is the Height of Activity for Nesting Yardbirds

MAY 28, 1995

The spring migration is essential over, and it is now time to pay attention to some of the nesting yardbirds. Mornings are best. That is when the birds, especially male birds, are most vocal and tend to move about more in "posting" their territories and searching for food. Exceptions are the nocturnal species such as owls and nightjars. May in South Texas is when bird activity is at its peak.

At my home in Mission Oaks, near Mission Valley, I can record a number of bird species by simply sitting out on my deck or taking a short morning walk. The most vocal of all my yardbirds is the Northern Mockingbird or "mocker." One of these gray, white, and black birds has claimed my mailbox and that of my neighbor's as singing posts. And woe be to any other mocker or, for that matter, any other bird that ventures nearby.

Yesterday my mocker actually chased away a Red-shouldered Hawk, a predator that is eight to ten times larger. When the hawk perched on an oak limb in my yard, the mocker actually struck it three times from behind. The hawk simply ruffled its feathers and tried unsuccessfully to ignore the pesky mockingbird. It eventually flew off to another site where it would not have to put up with such abuse.

Other common birds in my yard at this time of year, listed in general order of abundance, include Northern Cardinal, Carolina Chickadee, Tufted Titmouse, Ruby-throated Hummingbird, Inca Dove, Brown-headed Cowbird, Painted Bunting, Buff-bellied Hummingbird, Blue Jay, Carolina Wren, and White-eyed Vireo. Somewhat less abundant species include the Yellow-billed Cuckoo, Ladder-backed and Red-bellied Woodpeckers, Mourning Dove, Common Ground-Dove, Greater Roadrunner, Great Crested Flycatcher, Lark Sparrow, Common Grackle, Bronzed Cowbird, and House Sparrow.

There also are a number of aerial visitors, those birds that buzz the yard regularly. These include the Black and Turkey Vultures, Chimney Swift, Barn Swallow, and Purple Martin. Flocks of Cattle Egrets pass by in early mornings and late evenings. Less numerous aerial visitors are the Great Blue Heron, Great Egret, Crested Caracara, Red-tailed Hawk, and Cliff Swallow. During dawn and dusk, Common Nighthawks can be seen and heard. And both Barred and Great Horned Owls can often be heard after dark.

Over the last five years, since I moved into my Mission Oaks home, I have recorded a total of 137 species of yardbirds. I have had great fun and considerable pleasure attracting and admiring our feathered friends.

"Welcome Home" Provided by Snakes

MAY 31, 1998

After three weeks in Big Bend National Park, where I led two seminars and a World Wildlife Fund tour (and encountered only four snakes in all that time), on arriving at my home near Mission Valley I found two snakes almost on my doorstep.

Most folks would consider it a rather unwelcome homecoming, but it was most suitable for me. It reinforced my respect for our part of the world, one that contains an amazing diversity of wildlife. Both snakes found on my doorstep were the nonpoisonous types, although the poisonous copperhead, often found in the surrounding ground cover, is usually the species most often seen around the house.

The first snake I encountered was a bright green-colored Rough Green Snake, about 20 to 24 inches in length and hardly more than finger-width. It didn't even try to crawl away, and I easily picked it up and examined it. But like several of the docile snake species, it immediately defecated, behavior that often makes predators drop their prey, and then the snake tried to escape. It was a truly beautiful snake, grass-green with yellow color on its lips, chin, and belly.

Rough Green Snakes are arboreal species that spend most of their time in trees, where they hunt various lizards and insects, including grasshoppers, crickets, dragonflies, and the like. They are far more common than the few records might suggest, because their color offers wonderful camouflage among the green leaves. They seldom come to the ground.

The second snake I encountered was a Plains Blind Snake, a light brown (the dullest of colors), extremely small snake. They rarely are more than 10 inches in length and only half the width of a lead pencil. This tiny creature was only 5 inches in length, and I probably would have missed it if I had not bent down to retrieve an object that I had dropped. In fact, when I showed it to my wife, she thought that it was only an earthworm. Some earthworms are larger than the Plains Blind Snake.

The name "Blind Snake" is rather confusing because this little creature does have tiny eye vestiges that look like black dots. However, its rounded tail strongly resembles its head, a characteristic that can aid in its escape from a predator not knowing which end to attack. And instead of living in green foliage like Green Snakes, Blind Snakes live underground, where they prey on termites and larvae and pupae of ants. Unlike the Green Snake, they coat themselves with feces, a clear liquid that repels insects. They are thus able to utilize ant tunnels and ant nests without being attacked by the ants. They also possess a unique feature in that their blunt tail contains a tiny spine that they use to anchor themselves to soil, helping them to move up and back through the soil and narrow tunnels.

Blind Snakes occur only in the southwestern portion of the United States, while Green Snakes are more widespread throughout the southeastern quarter of the country. Three forms of Blind Snakes, all species of the genus *Leptotyphlops* (pronounced lepto-tie-flops) occur in the United States: Trans-

Pecos Blind Snakes occur along the Rio Grande from near Del Rio west through far West Texas to southern California; New Mexico Blind Snakes occur in West Texas and north to central Oklahoma; Plains Blind Snakes occur in central Texas to the Coastal Bend and south into central Tamaulipas, Mexico. They all look alike and are seldom reported.

Finding one of these little snakes or one of the colorful Rough Green Snakes helps one appreciate our wildlife heritage even more.

June

"Rain Crow" Cuckoo Nests in South Texas

JUNE 1, 1997

The very last of our nesting birds—the Yellow-billed Cuckoo—arrives on its nesting grounds in South Texas during the last of April and the first week of May. By the first of June it is fairly common and most evident by its distinct song, "kakakaka-kow-kow-kowp-kowp-kowp," a guttural sound that starts fast and slows at the end. This long-tailed bird also makes its presence known by flying across openings from one tree to another. When seen up close, our cuckoo is about 12 inches long, including a 6-inch tail with large white spots on a black background below, all white underparts, rufous wing-patches, and a slightly curved yellow bill.

Yellow-billed Cuckoos spend their winter months in South America, from northern Columbia to southern Brazil, but they leave the tropical forests for more northerly nesting grounds from northern Mexico throughout the southeastern two-thirds of the United States. They nest in second-growth woodlands, forest edges, thickets, orchards, and such, including yards with broadleaf trees. Their nests are little more than a few sticks placed on a branch or in a tree crevice, and their three or four pale greenish eggs produce young in about fourteen days. The adults and young remain for several more weeks and normally leave South Texas by late September.

Many folks know our Yellow-billed Cuckoo best by the name "rain crow," due to its habit of calling frequently on cloudy days or just prior to rain. Its taxonomic name is *Coccyzus americanus; Coccyzus* is Latin for "to cry cuckoo," referring to the birds' strange song. It also is known for its habit of consuming large quantities of caterpillars, some of which are injurious insect pests. It isn't an exception to find a Yellow-billed Cuckoo tearing apart tent caterpillar webs to capture the caterpillars. It also feeds on a wide variety of beetles, grasshoppers, ants, wasps, flies, and crickets as well as various fruits such as wild grapes and mulberries.

Our cuckoo is a member of a huge family (Cuculidae) that includes 129 species worldwide. North American relatives are limited to the Black-billed Cuckoo that migrates through our area to its nesting grounds in the northeastern states, Greater Roadrunner of the Southwest, Groove-billed Ani of extreme South Texas, and the Mangrove Cuckoo and Smooth-billed Ani that occur in South Florida.

Perhaps the best known cuckoo is the European Cuckoo that sings the typical "coo-koo" songs, after which the cuckoo clock was developed. That bird, along with several other European and African cuckoos, is a nest parasite, like our Brown-headed and Bronzed Cowbirds. They lay their eggs in the nests of other, smaller birds, and the foster parents raise the young cuckoos. This method of reproduction is one that takes very little energy by the true parents but consumes considerable energy by the host parents. Our cowbirds have been extremely successful with this method, increasing their range at every opportunity. They have become true pests and are partly blamed for declining populations of several of our favorite songbirds.

Although there are a few records of Yellow-billed Cuckoos also practicing nest-parasitism, that behavior is extremely rare for this species. It appears that our cuckoo is beneficial not just to the nature-lover but to the farmer and rancher as well.

Gaggle of Names Used for Animals and Their Young

JUNE 2, 1996

The names of young animals and groups of animals are not only poorly known, but too often are used incorrectly. Although almost everyone knows that the young of a bull and cow are calves, did you know that their young also are called "veals," "vealers," "stirks," and even "hogs," without reference to their sex? And male calves are called "bullocks," "stots," "bulls," and "bull-calves," and the females are called "heifers"?

The young of rabbits, hares, skunks, beavers, otters, ocelots, mountain lions, and bobcats, as well as house cats, are called "kittens." And everyone should be familiar with the words "pups" and "fawns" for young dogs and deer, respectively. But how many of you know what a squealer is? It is a young quail, not a pig!

When you hear someone mention "hens," you usually think of the female chicken, but the term "hen" also includes the females of fish and lobsters, as well as the female canary. And although the words "bull" and "cow" may refer to cattle, they also may refer to the male and female moose, terrapin, and several other animals.

Here are some additional names of young animals: antling of an ant, spiderling of a spider, cygnet of a swan, chicken of a turtle, cub of a fox, calf of a giraffe, gosling of a goose, chigger of a mite, maggot of a fly, and squab of a dove.

And then there are the names for groups of animals that are misused almost as often. Some of these, such as a school of fish, swarm of bees, pride of lions, and skein of geese, are reasonably well known; others are not. Did you know that a "gaggle of geese" is the proper terminology when the geese are on the water? And groups of swallows are known as flights. But what are gulps, murders, dules, budlings, and charms? It would be proper to refer to a gulp of cormorants, a murder of crows, a dule of doves, a budling of ducks, and a charm of finches. Groups of hawks, herons, magpies, and owls are properly known as a cast (hawks), siege (herons), tiding (magpies), and parliament (owls).

Here are some additional group names: a covey of partridge, nye of pheasants, host of sparrows, wisp of snipe, masting of storks, spring of teal, rafter of turkeys, pitying of turtle-doves, fall of woodcocks, and decent of woodpeckers. And groups of wolves are properly known as routes, groups of squirrels as deays, turtles as bales, and toads as knots.

Numerous other odd names are used for animals, many of which are used only by those individuals with special interests. But the English language is sprinkled with fascinating names for animal young and groups, although they may be seldom used.

Watch Out for Copperheads

JUNE 4, 1995

Summertime, with its hot and humid conditions, produces just the kind of environment that is favored by Copperheads. Anyone working outside, gardeners and others alike, should be especially watchful for these poisonous reptiles. Although this snake is most likely to occur in thickets, along streams, and in moist rocky places, the ground cover around homes in the country seems to be another favorite habitat.

Copperheads are responsible for the majority of all poisonous snakebites suffered by humans throughout the southern states. Although this snake is relatively docile and normally will strike only when provoked, its habits can easily conflict with our own activities. However, human fatalities are rare, and serious problems seldom arise when a bite is properly treated. According to Alan Tennant in his book, *Field Guide to Texas Snakes,* "not a single death resulted from 308 copperhead bites over a 10-year period" in Texas. Tennant claims that Copperhead poison is only half as destructive as that of a Western Diamondback Rattlesnake.

During summer, Copperheads normally spend their days hidden from view in heavy vegetation and are active only at night. During cooler weather they may be more active during the daylight hours. Prey species include a wide range of small mammals, lizards, frogs, and insects. Tennant points out that White-footed and Harvest Mice are probably their principal food species. Anoles and geckos are also readily available and are additional prey species commonly used.

Broad-banded Copperheads are surprisingly common in South Texas.

The Copperhead, known to scientists as *Agkistrodon contortrix*, is a rather stout snake that is rarely more than a yard long; the largest one recorded was 4 feet 5 inches. Copperheads are easily identified by their rich coppery brown color and thirteen to twenty darker brown crossbands over the back. Their heads are also rather distinct. The head, where its recurved, movable fangs are located, is not only wider than the neck, a characteristic typical of all pit vipers, but also sports an elongated pinkish brown patch along and above each jaw, as well as vertically elliptical pupils.

A heat-sensor pit is on each side of the head a little below and behind the snake's nostrils. This sensory organ acts as a heat receptor to detect prey and help the snake aim when striking at warm-blooded prey. Herpetologist Roger Conant, in his classic *A Field Guide to Reptiles and Amphibians of Eastern and Central North America*, warns that for freshly killed specimens, "Reflex action may last a long time, and supposedly dead pit vipers have been known to bite."

Mating occurs in early spring, soon after the snake emerges from hibernation. The female carries developing eggs inside her body all summer, and young are born (5–6 per litter) alive in late summer or early fall (August and September). However, the young Copperheads, roughly 9 inches long, are born in sacs, the last relics of egghood. In half an hour, already poisonous enough to kill small prey, they will break from their sheaths and strike out on their own.

Camel Crickets Are Commonplace

JUNE 7, 1998

It is again that time of year when Camel Crickets can be found in innumerable dark places around the outside of the house. They also seem to prefer places inside the door screens or garages where they can venture out after dark in search of food.

My deck is a favorite location this year, where they can come up through the cracks to forage. And if one happens to enter the house, it will readily seek a dark corner to hide in before I am able to capture it. What bothersome but harmless creatures they are!

Camel Crickets are very different from the Field Crickets that can be abundant later in the summer. Camel Crickets look somewhat like a camel due to their arched back. They also have been described as "shrimp with big legs." They are big bugs, an inch or more, with overlapping plates that, shrimp-like, armor their backs. They are usually tan to gray-brown in color and possess very long antennae, arched over their back, and long hind legs for jumping. They are great jumpers, but they also spend much time walking about.

Camel Crickets, however, are not true crickets. They actually are more closely related to the long-horned wingless grasshoppers. Yet unlike the grasshoppers, they prefer nighttime over bright sunlight. And due to their nocturnal habits they are often called "cave cickets," which also is not a valid name. True cave crickets are usually blind and lack any pigment, features derived from living underground. In the wild, Camel Crickets are usually found in twilight locations, such as in hollow trees and beneath logs and stones.

A member of the Orthoptera family, a large group of insects that includes grasshoppers, crickets, katydids, cockroaches, mantids, and walking sticks, Camel Crickets differ from many of the orthoptera in several ways. The most important difference is their inability to sing, like most of the other orthoptera. True crickets are famous for this ability, and some species are even kept in cages so their owners can enjoy their songs. Camel Crickets, however, lack any auditory function, being deaf to airborne sound.

The vast majority of their sensory perception comes through their long, graceful, sweeping antennae. They use them to know their world, to feel their way about, as well to communicate with other Camel Crickets. Researchers have reported, when keeping several Camel Crickets in a terrarium, that they often form circles with their antennae touching. And when they are fed, they often shove one another about to obtain the food but rarely loose touch with their antennae.

Researchers also report that these crickets spend considerable time grooming themselves. They clean their antennae by pulling them through their mouths like a man chewing his mustache, working along the length of them until the tip end snaps free and they pop back into place. The males even bend double under themselves and groom the tips of their abdomen.

Camel Cricket food varies, but it primarily consists of animal debris of all sorts. Their diet primarily consists of dead flies, spiders, moths, and even cricket corpses. Researchers maintain populations by feeding them crushed dog food. Bottle caps make good feeding and watering trays. They do not feed on plants, as many other orthoptera do. They, therefore, are not a problem to crops and other plant life. If anything, they are a benefit to our environment. And for those individuals looking for a unique pet, how about keeping Camel Crickets? That would be a very different hobby!

Killdeer Is Often a Farmer's Best Friend

JUNE 9, 1996

The other day I discovered a baby Killdeer in the middle of the driveway of my bank in Victoria. I marveled at its small size, compared with the adult that was standing nearby, trying to keep its baby from being run over. Yet this tiny creature had all the same features of the parent, double black breast band against its all white underparts, black-and-white face pattern, buff back, reddish orange rump, and yellowish legs. Except for its size and obvious stupidity, it was just a Killdeer in miniature.

The Killdeer is a bird that we too often take for granted because it is so widespread and common, but it truly is one of our most amazing birds. First of all, it has adapted to situations in which few other birds could survive: it is able to nest on our lawns, in pastures, gravel pits, mud flats, dry and relatively barren flats, and even on rooftops. It often nests far from water. Killdeer plumage seems too contrasting for good camouflage, yet, like a zebra, its bands break up the bird's outline against the background, providing it with what is known as *disruptive coloration*, which is excellent protection.

Killdeers also have developed a remarkable distraction display, the behavior of pretending injury, usually running off and dragging a wing like it is broken to entice predators away from a nest. But just as it is about to be caught it will suddenly recover and fly away. And finally, its babies are so precocial that they leave the nest within a few hours of hatching and run about with their parents.

The adaptive Killdeer, actually a plover of the shorebird order, occurs from Alaska south to South America and has even been found in Hawaii, Greenland, and Great Britain. Northern birds migrate south for the winter, but most of the South Texas Killdeers are year-round residents. Yet we also find flocks of migrants from mid-March through May and again in September and October. Then and all the rest of the year they feed on insects, especially beetles, grasshoppers, and dragonflies; they also will consume other small animals. They are considered one of the farmer's best friends.

But for all its adaptability, its friendliness, and its helpfulness, its best known characteristic is its voice. Although it has a variety of calls, from plaintive "dee-ee" notes to thin trills of "tit-tit-titititit," its typical "kill-dee" call, from which its name was derived, is most familiar. And, when disturbed, it may carry on for an amazingly long time, even to the point when it becomes exasperating. It has certainly earned its name: "kill-dee, kill-dee, kill-dee."

Crested Caracara Families Are Now Out and About

JUNE 11, 1995

Family groups of Crested Caracaras, two adults and two or three fledglings, are now flying over the fields and brushland of South Texas. The adults are teaching their youngsters the art of caracara-predation and which of the abundant roadkills and other carrion to utilize. This stage in the young caracara's training is vitally important for its long-term survival.

To many South Texans, Crested Caracaras are best known as "Mexican eagle" or "Mexican buzzard," names derived from Mexico, where it is that country's national bird and depicted on its colorful flag. These names are somewhat appropriate because the bird does resemble an eagle (or a Common Raven) in flight and often feeds on carrion with buzzards, namely the Black and Turkey Vultures. Yet it is not closely related to either of these birds.

Actually, the caracara is a falcon, more closely related to Peregrines and American Kestrels. But unlike those two falcons, it is a bird of the open brushland of South Texas and Arizona (there also is a population in South Florida) and southward to northern South America.

According to Harry Oberholser's *The Bird Life of Texas*, its name is of "Guarani Indian origin, derived from the bird's infrequent 'cara-cara' cry (also rendered 'traro-traro')."

The caracara is truly an outstanding bird. Not only is it a large bird—hawk sized—but it sports contrasting plumage: black back and cap, with contrasting white cheeks, throat, tail, and noticeable wing-patches in flight; a streaked or barred breast; and a yellow face and long yellowish legs. It has a loud call when disturbed, a grating "trak-trak-trak," like a stick drawn rapidly across a wooden washboard, or a bird clearing its throat.

Although ornithologists generally agree that the caracara is a true falcon, it is far more adaptable than most other falcons. It is known to nest on trees, on giant cacti in the Sonoran Desert, on cliffs, and in South Texas on yuccas, large shrubs, and in oak mottes. However, there is no better example of its adaptability than its feeding habits. One time it may be seen perched on a dead cow or deer carcass, feeding with vultures along the roadside, while the next time it may be found attacking prey from the air, falconlike. In addition, it also is able to run on the open ground, chasing down lizards, snakes, and small rodents.

Arthur Cleveland Bent, in his Life History series, lists the following foods: "rabbits, skunks, prairie dogs, opossums, rats, mice, squirrels, snakes, frogs, lizards, young alligators, turtles, crabs, crayfish, fish, young birds, beetles, grasshoppers, maggots, and worms." Bent also points out that in Texas, caracaras will also harass larger birds that are carrying food; when the food is finally dropped, they will scoop it off the ground or from the water for themselves.

What other bird is so adaptable in its feeding habits and has such a remarkable personality as our Crested Caracara?

Even Butterflies Are Affected by Our Hot, Dry Weather

JUNE 14, 1998

Butterfly numbers in South Texas are only a small percentage of what they were last year and even the year before. I am finding fewer species as well as fewer individuals. Watering and maintaining a good variety of flowering plants in my yard helps to attract whatever species might be passing by, but the hot, dry conditions have affected even the most common species.

One can still expect a few species with fair certainty. The largest of these are two swallowtails: the Giant Swallowtail, a black butterfly with broad diagonal and marginal yellow bands, and the Pipevine Swallowtail, blackish with shiny blue at the trailing edge of its hindwings. Another large and showy butterfly that remains fairly common is the Gulf Fritillary, a bright orange creature with long wings with scattered black dots; the underside shows black-rimmed silvery blotches.

The frequently sighted sulphurs are limited to Large Orange Sulphurs and Little Yellows. The Large Orange Sulphur (wingspan of almost 3 inches) is an apricot-orange color, although some females and faded individuals can be almost whitish or peach colored. All, however, show a distinct dark line running down the underside from the tip of its front wings and scattered reddish spots. The other common yellow butterfly is the Little Yellow or Little Sulphur, depending on your field guide. It is also lemon yellow but with a wingspan of only about 1 inch, and it too shows a narrow black margin and tiny black dots.

And then there is the common Checkered-Skipper that seems to be about as abundant this year as last. It is the same size as the Little Yellow but is checkered black-and-white, with bluish body hairs. The much larger Monarch look-alike that is present in drier areas is the Queen Butterfly. It is smaller than a Monarch without the Monarch's obvious black veins; it does possess black margins with tiny white spots.

Two additional butterflies are especially numerous in shady places in my yard: Dusky-blue Groundstreak and Carolina (or Hermes) Satyr. The Groundstreak is a very small (about 1 inch) tan-colored creature that possesses narrow red lines on its front wings and hindwings and a black tail spot with an orange cap; it takes a good close look to see this, but it is well worth the effort. And the Carolina Satyr is all brown above but with a series of large and small eyespots on the underside. It flits about near the ground, and it requires patience to see it well.

I have also seen an assortment of other species during the last month but in very small numbers. These include Cloudless and Orange Sulphurs; Gray Hairstreak; American Snout; Silvery Checkerspot; Texan and Pearl Crescents; Question Mark; Common Buckeye; Goatweed Leafwing; Little Wood-Satyr; Gemmed Satyr; Sickle-winged, Clouded, Fiery, Dun, and Eufala Skippers; and Celia's Roadside-Skipper.

How to attract more butterflies? A variety of flowering plants and moist conditions is usually the key. I also have placed out fruit, such as over-ripe bananas, that seems to work, if those species—especially the Emperors—are in the neighborhood. Some folks have also tried putting out a "sugar mix" as an attractant. I have had only minimum success with this, but in case you are interested, here is the recipe: 1 can of beer, 1 over-ripe banana (skin included), 2 teaspoons of black molasses, 1 tablespoon of fruit juice, and 1 pound of sugar. First, dissolve the sugar in the beer, then mash in the banana, add the molasses and juice, and heat this mess for 15 to 20 minutes, stirring gently until it begins to boil. Then reduce the heat and let it simmer. After about 10 minutes, allow it to cool, stirring every few minutes, and then bottle while lukewarm. That goop can be dripped onto a feeding tray or painted (with a paintbrush) on any surface.

The Crested Caracara is a falcon, but often acts like a vulture.

Some folks find that this mess works very well for both butterflies and moths. Good luck!

Click Beetles Are Amazing Creatures

JUNE 21, 1998

Finding a giant click beetle recently reminded me of those that I played with as a youngster. I would place one of these hapless creatures on its back and wait. In a short time—once, I assume, the beetle thought it was safe—it would suddenly throw itself in the air with an audible "click" and land on its feet. If it did not land on its feet and happened to land on its back, it would repeat this process time and again until it did. Then it would crawl away to safety. But as a youngster, the game was to make it repeat its performance many times, until I tired of that amusement. I imagine that the beetle got tired of throwing itself up in the air before I did.

Click beetles are amazing creatures! How the behavior of righting itself by clicking itself into the air evolved is anyone's guess, but it apparently has worked for many generations of click beetle, of which there are several hundred species worldwide. The process is necessary because the insect's short legs are not long enough to help it right itself after falling on its back, a shortcoming that allows some curious naturalists (no matter the age) to experience such odd behavior for themselves.

The click beetle apparatus consists of a spine on the underside of its thorax which slides in a groove below the mesothorax (the middle portion of the thorax). While lying on its back, the beetle straightens out, bending the thorax farther and farther back, as in cocking a gun, until the spine is at the very end of the groove. Suddenly, slipping out of the groove, there is a snap, with an audible click. The release of the muscular tension forces the shoulders or base of the wing covers against the ground with such force that the insect is thrown into the air, spinning end over end, sometimes to the height of 4 or 5 inches.

My click beetle is known as an Eyed Click Beetle and is one of the largest (almost 2 inches) of the family of Elateridae, a group also known as "skip-jacks." Scientists know it as *Alaus oculatus.* The "eyed" term comes from the fact that it possesses two rather large, oblong, black "eyes" on the top of its head. These are not true eyes but only features that the species has developed as a protective measure. The two huge eyespots make the click beetle appear all the world like some weird creature with an extra set of eyes. Predators are certain to take a second look. Tiny true eyes are located on front of the head, where they should be, but they are hardly noticeable. The remainder of the elongated, hard-bodied beetle is blackish with grayish blotches.

Elater beetles live under bark and on vegetation, and the adults do little damage. The larvae of some species, however, commonly known as "wire-worms," are often destructive, feeding on newly planted seeds and the roots of beans, cotton, potatoes, corn, and cereals. Other larvae occur in rotting logs or in ant and termite nests, and these feed on other insects. Pupation occurs in the ground, under bark, or in dead wood.

Eyed Click Beetles are not one of the bad guys, but seem more like a curiosity and something to make a slow summer afternoon something special.

Chickadees Are Nesting for a Second Time

JUNE 22, 1997

The Carolina Chickadees that reside in my yard are already nesting for the second time this year. The first brood appeared about my feeders and birdbath in early April, and all those youngsters now are next to impossible to separate from their parents. They make up a lively and vocal troop of black-and-white birds, each with a personality of its own. Some seem to like to bathe daily or even several times a day, while others, perhaps the youngsters that have not yet learned to appreciate a good cool bath at midday, seem hesitant.

In spite of their diminutive size, chickadees are among our best loved songbirds. From the abundance of chickadee designs that appear on plates, pillows, and all sorts of things, even nonbirders seem to appreciate chickadees. One reason that chickadees are so popular is their widespread distribution across North America. In fact, six species of chickadees occur regularly within the continental United States. Our Carolina Chickadee is the most widespread, residing throughout the eastern forests from Maine to Illinois and south to central Florida and southwest to central Texas. The San Antonio River marks the southernmost edge of its range.

Other North American chickadees include the Black-capped Chickadee, which is common all across the northern half of the continent but which rarely gets into Texas; some winters one or a few may reach the upper Texas Panhandle. The Mountain Chickadee is a true mountain bird that occurs from the Yukon south to New Mexico; it occurs in Texas only in the forested areas of the Guadalupe and Davis Mountains. The Chestnut-backed Chickadee is found only along the West Coast from Alaska south to Southern California. The two other resident chickadees occur within very limited areas: the Mexican Chickadee, common throughout the mountains of Mexico, reaches into the United States only to the mountains of southeastern Arizona and southwestern New Mexico; the Boreal Chickadee, although fairly common throughout Canada and most of Alaska, occurs in the rest of the United States only in Maine and rarely along the Canadian border in Montana, Wisconsin, and Michigan.

Although all these little birds appear alike at first glance, they possess very distinct songs, calls, and plumage. Our Carolina Chickadee whistles a fast "chick-a-dee-dee-dee" song, while the Black-cap sings a lower, slower version. The Mountain Chickadee sings a descending four-note whistle, like "chick-adee-adee-adee-adee." Chestnut-backs sing a hoarse but rapid "tseek-a-dee-dee;" Mexican Chickadees possess a warbled whistle; Boreal Chickadees sing a slow, nasal "tseek-a-day-day."

Except for the Carolina and Black-capped Chickadees, most chickadees can easily be told apart by rather distinct features. Mountain Chickadees possess a white line between the black cap and throat; Chestnut-backs possess an obvious chestnut back and flanks; Boreals have a brown cap and flanks; Mexican Chickadees possess a relatively short tail and gray flanks. Separating the Carolina and Black-cap can be difficult, especially if they are not vocalizing or are in worn late-summer plumage. The best identifying features are their wings; Black-capped wings appear two-toned, possessing a broad whitish or milky wash that extends from the shoulder to near the wing tip. Carolina Chickadee wings are a solid blackish color.

Our abundant Carolina Chickadee is one of our most personable songbirds, living in our yards year-round; they do not migrate. Pairs form in the fall as new families gather together in flocks that usually remain together most of the winter, feeding with Tufted Titmice and a number of wintering birds, such as Ruby-crowned Kinglets and Yellow-rumped Warblers. By late winter the flocks break up, and pairs establish territories that they keep throughout the nesting season. A cavity nester, they utilize a wide variety of natural holes in trees and shrubs, as well as deserted woodpecker cavities and human-made structures such as bluebird boxes and the like. The female selects the site, and both parents feed the nestlings; the young are fledged in thirteen to seventeen days. A normal family will consist of five to eight youngsters.

Usually post-nesting birds will join other families, and unless a pair nests for a second time, those mixed flocks are likely to stay together until the following nesting season.

The Beautiful But Gaudy Painted Bunting

JUNE 23, 1996

No other bird can be confused with the male Painted Bunting. This colorful little bird is almost unreal. To me, it is gaudy, but my wife thinks it is the most beautiful of all birds. Males possess a deep blue head; green (almost chartreuse) back; bright red underparts, rump, and lower back; brownish wings and tail; and red eye ring. Whether it is considered beautiful or gaudy, when seen up close, it can hardly be ignored.

The sparrow-sized Painted Bunting is one of our neotropical migrants that arrives in South Texas in early spring, It immediately establishes its territory and nests, and its young usually are fledged and on their own by mid-June. The female, a plain, greenish yellow bird, does most of the household chores, while the colorful male is defending its territory. Most of this involves singing from various posts; it will vigorously chase away other Painted Bunting males when necessary. Its song is a surprisingly loud (for such a small bird) and clear, musical warble, like "pew-eata, pew-eata, j-eaty-you-too."

Males seldom remain all summer and usually head south to their wintering grounds (Mexico to Panama) by late July. Females and the youngsters often remain until September. While in South Texas, males and females will readily come to seed-feeders; I have two or three males and six to eight females at my feeders daily. And the males will often sing their sweet songs while perched in the tree just above the feeders.

There is a very good reason why we often see more females than males at the feeders. Painted Buntings, like many other birds, are polygamous; a single male mates with several females. Although humans tend to question this practice, it is a very practical one for many birds, particularly for neotropical migrants that must select a territory, court, mate, and raise a family in a relatively short period of time. The females simply choose the male that is best able to claim and hold the most superior territory. A resource-rich territory will most likely provide a better chance of producing offspring than an inferior one.

With this better understanding of the Painted Bunting lifestyle, you probably will never again look at one of these amazing creatures with the same perspective.

Five Kinds of Doves Are Native to South Texas

JUNE 25, 1995

The mournful songs of doves are some of our most memorable sounds in spring and early summer when they are defending their territories and courting their mates.

Like most birds, male doves spend considerable time marking the edges of their territories, singing from prominent posts or, when necessary, from the ground. It is their way of telling their competitors that the area is claimed and of reassuring their mates that all is safe and secure. Their actions follow their ancestral behavior that is part of their very nature, imprinted on their genes over the millennia.

In South Texas, we are fortunate to have five native species of doves, while the majority of the United States and Canada claim only one—the Mourning Dove. We also have the White-winged, Inca, and Common Ground-Dove in the north, and from Goliad and Refugio counties south, the White-tipped Dove.

Each of the five species sings its own unique love song, and each requires slightly different needs. But all are seed- and fruit-eaters, although insects also form a large part of their diet, especially for the nestlings.

The five doves are fairly easy to tell apart. The largest of the five is the White-tipped Dove, a rather plain dove with noticeable (in flight) white tips on the tail, and the White-winged Dove with large white wing-patches. The White-tipped Dove sings a very deep song that sounds like someone blowing across the mouth of a bottle. The White-winged Dove sings a rolling "Who-cooks for who, who-cooks-for-whom."

Almost everyone knows the song of the Mourning Dove; it is one of the most common of our summer birdsongs: "Who-ah, whoo-whoo-who," with a rising second syllable. This common species is uniform gray-brown with a small head, blue eye-patches during the breeding season, and a long, tapered tail with white feather-tips. Mourning Doves usually congregate in flocks after nesting, and in fall they make exciting game birds.

The two smallest doves prefer very different surroundings. The Inca Dove, with its scaly plumage, black bill, long, narrow tail with white borders, and reddish wing-patches in flight, frequents our yards and gardens. The even smaller Common Ground-Dove prefers open, more arid areas. It also sports rufous wing-patches but possesses a short, dark, rounded tail, light yellow bill with a black tip, and grayish brown plumage with black spots.

Inca Doves have a variety of songs and calls but usually sing a hollow and repetitious "whirl-pool," with an accent on the second syllable. The Common Ground-Dove sings a low repetitious "who-oo, woo-oo, woo-oo, woo-oo," with a rising inflection at the end of each syllable.

The opportunity to enjoy five species of doves rather than only one is another bit of evidence in support of the varied environments of South Texas. It is further proof that we live in a marvelous portion of the world, which we must commit ourselves to preserve each time we hear the sad songs of the doves.

Seen Any Horned Toads Lately?

JUNE 27, 1999

Horned Toads, or Texas Horned Lizards to biologists, once were commonplace throughout much of Texas. They could be found in sandy areas throughout the state, but today these curious creatures have declined to a point where it is extra special to find one. In fact, the species was listed as "threatened" in Texas as early as 1977.

The Horned Lizard decline has been blamed on insecticide use, overcollecting for the pet trade, and the increase of the imported fire ants. Although all of these reasons may have contributed to the decline of Horned Lizards, herpetologists (scientists who study reptiles and amphibians) from Texas A&M University–Kingsville recently claimed that the Horned Lizard decline may also be due to malaria. The researchers have found many individuals suffering from this mosquito-borne disease, and they plan to continue their surveys throughout the summer months to test their theory.

If malaria is truly the reason for the widespread decline of these fascinating lizards, it is just as likely that certain individuals will resist malaria and survive. Those more resistant individuals will eventually repopulate the state.

Horned Lizards are fascinating creatures for several reasons. One reason is their apparent indifference to human activity. They are easily captured, and they can often be maintained as pets. They survive very well on mealworms or some other similar food. In nature, they eat lots of ants. Another reason for their appeal is their general appearance, like a horned toad with a tail. The Texas Horned Lizard possesses a pair of elongated, scaly horns on the back of its head, several smaller horns along the edge, and many short spikes scattered across its back. Usually brownish in color, others may be yellowish or reddish. And all have the flattened appearance. They look a little like a tiny dinosaur.

They also possess the ability to spit blood from pores near the eyes. Such behavior occurs only when they are highly agitated, like when they are grabbed and threatened. This special characteristic allows some individuals to escape from predators that might get a squirt of blood in the eye.

The range of the Texas Horned Lizard is limited to Texas and Oklahoma, parts of Kansas, New Mexico, and Arizona, and most of northeastern Mexico. It is a popular and fascinating lizard, and the loss of such a creature would impoverish us all.

Mockingbird Songs Provide Clues for Some Other Species

JUNE 28, 1998

Almost everyone has marveled at the Northern Mockingbird's amazing repertoire of songs. A territorial mocker usually will sing from dawn to dusk and even long into the night if he is especially amorous. Who hasn't been awakened on a moonlight night by a mockingbird pouring forth his songs?

Yet we humans too often use our own interpretation of why songbirds sing, assuming that a singing bird does so primarily to attract or to court a mate. Although bird songs certainly have that effect on occasions, the principal purpose of a bird's singing is to defend its territory. It is a signal that that particular site is taken and that interlopers are not only unwelcome, but will be vigorously driven away. It is a sacred site that will be defended to its holder's death.

But why do mockingbirds sing such a variety of songs? And how does that individual learn such a wide assortment of songs, sometimes of birds the mocker may never have encountered? That is the million-dollar question.

Although some mockingbirds are migratory—northern birds that move to warmer areas in winter—most of our South Texas summer mockers are resident, utilizing the same area year-round. And although it is possible that those resident birds may hear and immediately learn songs of migrants passing through our area, some researchers suggest that many songs are culturally transmitted, learned from ancestors, generation after generation. Young mockers get their initial introduction into mockingbird song as a nestling. But to take the idea of a genetic connection a step further, could that mean that its repertoire could even include songs of extinct species?

Most young birds only learn songs of their own species, and ornithologists estimate that about 85% of a mockingbird's singing is "uniquely mockingbird." But what about the other 15%? Those songs are derived from all types of sources, including other nearby birds, a human baby's cry, an engine, a whistle, and an amazing assortment of other sounds. They rarely imitate extensive sounds, but rather simplify a phrase by utilizing only pieces. Mockers often imitate cardinals but seldom if ever imitate the more detailed and extensive songs of wrens.

How long will a mocker remember a song? There is some evidence that they possess short-term memory—they will mimic the songs of birds passing through the area, and then stop singing those songs until the same species passes by the following year. This could be considered "cross-tutoring"; their memory must be triggered by hearing a song again before they can include it in their permanent repertoire.

Yet birds seem to remember certain songs better than others. For instance, sounds made by predators or competitors can readily be remembered from one year to the next. In one study, tapes played of different songs elicited immediate defensive stands whenever those songs included predators or competitors whose territorial disputes had been settled the previous year. This suggest a fairly well-developed memory capacity, far better than we humans often give them credit.

The Northern Mockingbird is unquestionably one of our most belligerent birds, chasing almost any other bird—fellow mockingbirds as well as most other species—out of its territory. This also includes house cats, dogs, and human beings, when the occasion arises. Perhaps this belligerence is the key to its vocal repertoire. Perhaps its territorial behavior has evolved to a degree that it mimics whatever displeases it. If singing mockingbird songs can drive out invading mockers, singing other songs might also drive out other potentially competitive bird species. And if that has worked for untold generations, perhaps its vast variety of songs are intended to drive away whatever else aggravates it.

I could well imagine singing loud rock music, sirens, or a number of obnoxious human voices that bother me. Maybe mockers have the right idea!

Bald Cypress Is in Full Summer Dress

JUNE 29, 1997

Few of our native conifers change so drastically from winter to summer as our magnificent Bald Cypress. Unlike most other conifers that are evergreen, Bald Cypress trees are deciduous, totally losing their leaves in winter and growing them back by summer.

These grand trees are commonplace along the Guadalupe and San Antonio Rivers, in old meanders, as well as in a few other wet places, throughout central and eastern Texas. Some individual trees reach 100 feet in height and 6 feet in diameter. And in places along the riverway, the Bald Cypress trunks flare at the base, producing beautiful buttresses and occasionally "knees" that may protrude above the waterline for several feet.

But for all the tree's size and characteristic trunk, it is the appearance of the foliage that is most like other conifers. The light green to yellowish leaves are alternate, featherlike, ½ to ¾ of an inch in length, and pointed at the tip. The foliage looks very much like that of the Douglas Fir of the Pacific Northwest. Spring flowers are 5 inches long and borne on drooping clusters of small male cones with a few female cones at the branch tips. Fruits mature in the fall and are 1-inch, wrinkled, rounded cones, reminding one of Sequoia cones.

Its nearest relative is the Montezuma Bald Cypress that is found naturally from the lower Rio Grande Valley south to Central America. Unlike Bald Cypress, Montezuma Bald Cypress is evergreen, keeping its foliage year-round, and it may be considerably larger. The national champion tree in Hidalgo County, although only 45 feet tall, has a trunk circumference of 222 inches and a crown spread of 74 feet. A Montezuma Bald Cypress in Oaxaca, Mexico, stands 164 feet tall and is estimated to be 2,000 years old.

Bald Cypress trees played an important role in early Texas. Because the wood is so durable and strong, trees were cut and utilized in the construction of docks and bridges as well as railroad ties, benches, and even building shingles. A Bald Cypress also makes a good ornamental because of its form, foliage, and fast growth. Fall color can also be appealing, although fallen leaves can require considerable cleanup.

To nature lovers, those of us who appreciate nature's unique environments, Bald Cypress habitats, especially swamp forests that occasionally exist in protected places, are most enduring. It was in these habitats in the deep southeastern United States where the now-extirpated Ivory-billed Woodpecker once existed. These special places are now identified by their abundance of Spanish moss and various orchid species. Other indicator trees and shrubs include Redbay, Tupelo, willow, sycamore, Box Elder, Buttonbush, Fox Grape, Poison-Sumac, and Carolina Rose. Bird life in these areas, today, usually includes Wood Duck, Northern Parula, Prothonotary Warbler, Pileated Woodpecker, White Ibis, and Barred Owl. American Alligators and Cottonmouths reside here as well.

Bald Cypress trees are no longer as abundant as they once were, not so much due to humans harvesting these marvelous trees, but because of increasing developments for housing and shopping centers, as well as enthusiasm for manicuring the environment.

America's freshwater wetlands are among our most endangered habitats! Our Bald Cypress swamps and riverways are not just special places of scenery and wildlife—they contain a truly unique ecosystem that humankind must protect if we are to maintain an intact life-support system.

Velvet Ant Is Really a Brightly Colored Wasp

JUNE 30, 1996

Hot summer days bring out a variety of insects that have been hiding in cracks and crevices, as well as those that have only recently hatched from eggs or metamorphosed from other forms.

Velvet Ants are beautiful wasps, not ants at all.

One huge group of insects is the Hymenoptera (meaning "membrane wings"), which contains ants, bees, wasps, and their relatives. All of these insects possess four membranous wings (the hind wings are smaller than the anterior wings and have a row of tiny hooks on the front by which the hind wings attach to the front wing), chewing or chewing-sucking mouth parts, antennae with ten or more segments, and a well-developed ovipositor. All go through complete metamorphosis; the larvae are grublike or maggotlike.

One of the most interesting—and confusing—of the Hymenoptera is the Velvet Ant. It is not an ant at all, but actually a brightly colored wasp, a member of the Mutillidae family. If you don't believe it, just try picking one up. They have a very nasty sting.

Velvet Ants come in a variety of sizes, from species that may be 1 inch or more in length to tiny ones that are about ⅛ inch. The giant of the family, which occurs throughout the southeastern United States, is called a "cow-killer ant," although it is far too small to kill a cow.

In my yard, I have seen at least three sizes of Velvet Ants, but the largest one is only about ½ inch. Actually, I have only been seeing the females that give them their "Velvet Ant" name. The females are wingless and densely covered with pubescence (fine short hairs) that usually is bright red, although others are banded with black and white or are all white. They also possess a velvety sheen. And, like ants, they run over the open ground, always in a hurry.

Male Velvet Ants are winged and usually larger than the females, but they also are covered with a dense pubescence. You can pick up a male, as they go about visiting flowers, without fear of a sting. The male and female are so different in appearance that they originally were considered two very different insects, at least until they were found mating.

The Velvet Ants' claim to fame is that they invade bee nests where they feed on the bees' larvae and pupae. Some prey on beetles and flies. But all the females possess a stinger that packs a real wallop!

July

Wood Storks Arrive for the Summer

JULY 2, 1995

Watching a flock of Wood Storks soaring high overhead, with the adult's all white body and coal black wingtips, is truly a beautiful experience.

On June 1, I found a flock of 65 Wood Storks soaring over the Guadalupe River near Nursery, and on June 16, Linda Valdez, Mark Elwonger, and I watched about 80 individuals high above Berclair. The two flocks were gradually moving northward.

While in flight, Wood Storks are among our most graceful birds, soaring with their long necks and legs stretched far out and looking surprisingly like White Pelicans, but up close they appear rather clumsy and ugly, at least by human standards.

One of our largest avian summer visitors, Wood Storks stand almost 4 feet high and possess a wingspan of about 65 inches. The common Great Blue Heron stands 4 feet tall with a wingspan of about 84 inches. While Great Blues and the other herons and egrets are rather appealing on the ground or in water, Wood Storks lack their feathered heads and daintiness. Their dark, naked heads and necks give them a vulturelike appearance. In fact, vultures have recently been reclassified by ornithologists (based on DNA studies) as storks rather than Falconiformes, the order of hawks, eagles, and falcons.

Wood Storks are long-legged waders that are true storks (family Ciconiidae) and not related to our more common waders, such as those in the families of herons and egrets, ibises and spoonbills, flamingos, or cranes.

Although Wood Storks once nested in South Texas, the nearest known nesting colony today is in southern Veracruz, Mexico; they also are residents of Florida's Everglades. Nesting occurs early in the year, and the Mexican birds (adults and juveniles alike) often move north into Texas and Louisiana by early summer. My June sightings were undoubtedly birds en route to their summering grounds.

Adults are easily identified by their all-white plumage with black flight feathers and tail, a distinct white plume below their short tail, and a black bill; immature birds possess a dirty plumage and a yellow bill. Both have a naked, black, and scaly neck and head and a long, stout, slightly curved bill.

During the summer, our visiting Wood Storks most often are found at woodland or prairie ponds, flooded fields, and fresh and saltwater marshes, usually in family groups or small flocks. When feeding, they wade about in the water, groping with their partly opened bill, swallowing any living thing they touch. Their most common foods include a variety of "rough" fish, frogs, snakes, young alligators, and even insects.

Although no longer part of the Texas breeding avifauna, their presence illustrates the close relationship that Texas has with its neighbor to the south.

Zebra Longwing Flutters Its Stuff

JULY 6, 1997

Finding a Zebra in my yard was a first! No, it didn't have four legs and wasn't grazing on grass; it was hovering around my Passion Flower vine. It flew off several times but each time returned to that flowerless vine to examine it closely, and it even landed for a few seconds before flying off to wherever Zebra Longwing Butterflies go.

It was the first time that I had encountered this gorgeous butterfly, but it is so distinctly marked that it is impossible to mistake it for anything else. The Zebra Longwing, or just plain Zebra, according to some books, is coal black with bright yellow to orange streaks running across the upper surface of its long, narrow wings, which also show a series of dots across the hindwings, with a rosy pink patch at each tip. And it also has a distinct way of hovering—rapid fluttering—and moving about slowly, with considerable sailing and drifting without an obvious direction. When disturbed, however, it can dart quickly away to safety.

Zebra Longwings are tropical butterflies that are rare visitors to the central Gulf Coast from Mexico and the Lower Rio Grande Valley. They have been recorded as far north as Colorado and the Great Plains. In the Coastal Bend, they are found most commonly nectaring on Lantanas and Passion Flowers. Geyata Ajilvsgi, in her wonderful book, *Butterfly Gardening for the South,* points out that Zebra Longwings are "one of the few butterflies with the ability to use pollen as a food source. Gathering minute amounts of pollen on the knobby tips of the proboscis, the butterfly releases a drop of fluid to dissolve the pollen; the insect is then able to drink the liquid in the usual manner."

Zebra Longwings are considered to be among the most intelligent of all our butterflies. Males and females roost together in low shrubbery each evening, sometimes (especially in the Tropics) in clusters of 60 to 70 individuals. They learn the locations of good flower sources from older individuals. And in mating, the male is attracted to the female pupae by scent. According to Ajilvsgi, "just before the female is ready to emerge, the male opens her shell with his abdomen and mates with the still unreleased female. He then deposits a repellent pheromone on the tip of the female's abdomen, which repels other males and thereby prevents her from mating again."

The female, once she emerges and begins flying, lays a few eggs each day over a period of several weeks or months. She may deposit up to one thousand eggs, depending upon an adequate diet of nectar and pollen. The average life span of a Zebra Longwing is usually less than four months. During that time, they usually stay within a few hundred yards of their home territories. But each successive brood gradually moves northward, so that products of the Lower Rio Grande Valley breeders begin to appear along the central Gulf Coast in late June or July.

They are most welcome—at least in my yard, where I have planted several dozen shrubs and vines to attract butterflies of all sorts. My Zebra Longwing was yard butterfly species number 86!

Alligators Are on the Comeback Trail

JULY 7, 1996

Alligators, with the possible exception of rattlesnakes, are our best known reptile. The alligator is not only the largest, but also one of the least under-

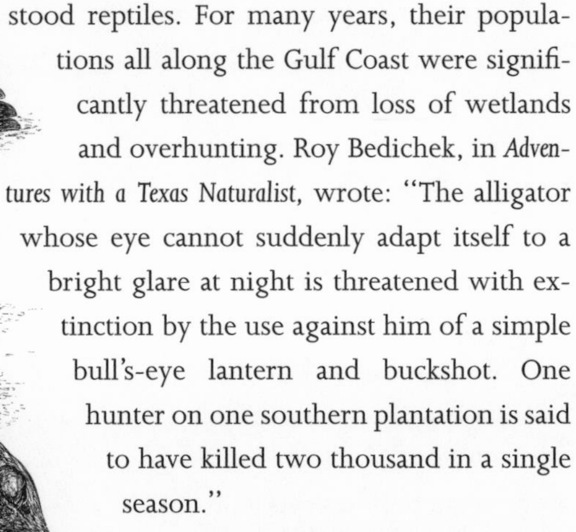

Alligators can reach 14 feet long.

stood reptiles. For many years, their populations all along the Gulf Coast were significantly threatened from loss of wetlands and overhunting. Roy Bedichek, in *Adventures with a Texas Naturalist*, wrote: "The alligator whose eye cannot suddenly adapt itself to a bright glare at night is threatened with extinction by the use against him of a simple bull's-eye lantern and buckshot. One hunter on one southern plantation is said to have killed two thousand in a single season."

Today, after several years of protection (it was given federal "endangered" status in 1967), it has recovered throughout its range and is almost as numerous today as it was in 1967. Texas held its first regulated alligator hunt in September 1984, and hunts have continued. Visits to numerous wetlands, from Lake Texana to Aransas National Wildlife Refuge, can easily turn up a few to several of these grizzled creatures. Wayne and Martha McAlister point out, in *A Naturalist's Guide: Aransas*, that the refuge population is estimated at 225 individuals.

Some alligators are huge, reaching 12 to 14 feet in length, but most average from 3 to 5 feet. The McAlisters explain that "Hatchlings are eight inches long. They grow about a foot per year for the first five years. Then the rate slows, but these reptiles continue to grow throughout life. An alligator matures when it is six feet long. Males grow faster than females: at an age of ten years males are about eight and a half feet long, females only seven feet. A twelve-foot bull alligator may be twenty years old and weigh 400 pounds; one massive wild-caught thirteen-footer tipped the scale at 850 pounds."

Most often, we find alligators quietly basking on an open bank, along a pond or stream, seemingly without a care in the world. When approached, they usually silently slip into the water and drift slowly away from any perceived danger. But when startled, they can move extremely fast, careening into the water with a huge splash that can give anyone nightmares.

Alligators are truly amazing creatures! They are one of our most primitive reptiles, having changed little since evolving even before dinosaurs. Yet they also are one of our most adaptable creatures. Although they prefer wetlands, they are able to travel considerable distance overland, and they also occasionally go to sea. The McAlisters write, "in 1965 a gator became entangled in a shrimper's trawl well out in San Antonio Bay." And what do they normally feed on? Almost anything smaller than they are, from crabs to frogs, fish, and anything else, on land or in the water, that they can catch.

Are they dangerous to humans? Attacks are extremely rare. The McAlisters state: "Generally a provoked gator will announce its disposition by inflating with air and hissing menacingly. A female guarding her nest or young will advance on an intruder, and she can move with unexpected speed. Beware of both ends; jaws chomp and tails swipe with leg-breaking force. So admire alligators from a prudent distance."

Chiggers Take A Bite out of Comfort

JULY 9, 1995

On a recent visit to a nearby cemetery in DeWitt County, my wife and seventeen family members got a full load of chiggers. Tiny red, itching welts were plentiful on their ankles, legs, and all around their waists the next morning—typical evidence of chiggers. I was saved from the same fate only because I wasn't present when they got infected. Chiggers have long been my nemesis; I can pick up those hateful creatures just walking down the driveway to fetch the morning newspaper.

There is a lot of misunderstanding about these tiny animals that are arachnids, not insects at all. My dictionary of biology states that arachnid is a "class" of arthropods that include scorpions, Whip-Scorpions, spiders, Harvest-men, ticks, and mites. They all possess six pairs of appendages: the first pair for grasping; second may be used for grasping, sensing, or locomotion; the remaining four pair for locomotion only; they do not have antennae.

Almost everyone recognizes true spiders as well as dog and deer ticks;

most of these arachnids are large enough to see reasonably well with the naked eye. But most mites and chiggers are another matter. Mites are extremely tiny arachnids with a body divided into two parts, the abdomen and the thorax. Mite mouths have piercing and grasping organs and a sucking beak. And the troublesome Harvest Mites, bright red creatures (sometimes called "red-bugs"), about the size of a pinhead, can sometimes be seen on the ground or in the soil.

Harvest Mites lay their eggs in soil or humus. On hatching, the larvae shed their skin in a few days and change into nymphs—the hateful chiggers! These nymphs are parasitic, requiring animal blood or fluids to survive. They climb onto plants to wait for a passing animal or human ankle. They then crawl into place, often passing through ordinary cloth because of their tiny size, to where they can feed on blood in skin pores or hair follicles, usually where the clothing is tight; they do not burrow into the skin. Severe itching and a rash often results and can persist for some time after the mites have left. Once they have fed, they drop off and soon molt into adult Harvest Mites.

Ticks and mites belong to the same order of animals, known to scientists as Acari. More than 15,000 species of Acari are known worldwide, and this is probably only a small fraction of what truly exists. Many are disease carriers and of economic importance, and so they have received considerable study. Anyone going into the field can usually avoid being attacked by using a good repellent. The itching can be eliminated or reduced by applying a weak Clorox and water solution or by taking a hot shower. And Wal-Mart pharmacist Javier Perez recommends two over-the-counter products, either Chigarid or Chiggerex.

Texas Armadillo, One of Our Strangest Creatures

JULY 10, 1994

The other evening I watched an armadillo in my backyard, slowly wandering this way and that in search of food. Its long snout was positioned at ground level so that it could detect anything edible. Suddenly it paused, stuck its muzzle in the soft earth and seemed to be eating whatever it had found. A few seconds later, it continued on, snuffling with its nose and grubbing with its forepaws, stopping every few minutes to dine.

As it came closer to where I was quietly watching, I could see its armored body, its powerful feet and claws, small leatherlike ears, and pinkish snout. What an incredible creature! When I raised my binoculars to get an even better look, it sensed my presence and immediately turned and ran off with a strange, rolling gait. Then, before it reached the wooded area beyond, it suddenly stopped, and, apparently realizing it was not being pursued, it continued to graze once again. I watched as it wandered off with its nose at ground level.

Although our armadillo is often taken for granted because of its abundance within South Texas, visitors from outside their range are often anxious to see one of these very odd creatures. And when the Spaniards first came to the New World, they too were amazed to find this "little armored thing," or armadillo. Although it does not occur in the Old World, it lives as far south as Argentina and has extended its range north throughout most of Texas and Oklahoma east to Alabama; it also was introduced into Florida where it is now common.

The Maya Indians believed that the Black Vulture turned into an armadillo in old age. The Mayans claimed that aging vultures gradually lost their wings and feathers. They would then enter a hole and start life anew as an armadillo. For proof of this story, they pointed out the similarity between the bald head of the vulture and that of the armadillo.

In many places, the armadillo is called "poverty pig" or "poor man's

The armadillo is one of our strangest mammals.

pig," as it sometimes is used for food. Its flesh is delicate, light, and tender, and when properly cooked, somewhat like pork in texture and taste. To scientists, our Texas armadillo is known as the Nine-banded Armadillo or *Dasypus novemcinctus*. It is the only member of the tropical family Dasypodidae (which includes 20 species) to occur north of Mexico. All the family members have protective "armor" of horny material, a soft and naked belly, and degenerate teeth; they have no incisors or canines but seven small peglike teeth on each side of each jaw. Undoubtedly because of its armor, armadillos are nonaggressive creatures that depend more on scent than sight for finding food.

Their diet consists mostly of ants, termites, and other insects, but they will eat a variety of small animals as well as carrion, some fruit, fungus, and a few other plants. William Davis, in *The Mammals of Texas*, reports that a study of 800 Texas armadillo stomachs revealed 488 different food items, 93 percent of which were animal matter: 28 percent was larval and adult beetles, 14 percent termites and ants, 8 percent caterpillars, and the remaining included earthworms, millipedes, centipedes, and crayfish. Bird eggs were found in only 5 of 281 stomachs.

Armadillos regularly visit water areas to drink and also to take mud baths. They can swim very well, although they ride low in the water due to their specific gravity. However, when necessary they can ingest air to inflate themselves to increase their buoyancy. But when small streams are encountered, they may simply walk across the bottom, emerging on the opposite side.

When not searching for food, armadillos utilize dens 2 to 15 feet long in rocky or soft terrain. Nest chambers, constructed at the end of the burrows, are 18 inches or more in diameter and stuffed with dried grasses, leaves, and other plant materials. They simply push themselves in and out each time they use the chamber.

Four youngsters are born from February to April, following a gestation period of about 150 days. The babies are identical to the adults without their hardened shells; the shell doesn't harden until they are almost full-grown. They very soon are able to follow mother like a flock of little piglets. She nurses them for nearly two months; she has four teats, so nobody goes hungry. Sexual maturity is achieved in the second year.

For all of the armadillo's oddities, its horny shell may be its strangest feature. It consists of many small plates of armor fitted closely together but able to slide on one another. Hard and stiff, the plates are jointed across the back so that the armadillo can curl itself into a hard tight ball with the shell on the outside and the head and feet tucked in out of harm's way. Few predators are able to get a grip on our Texas armadillo.

Eurasian Collared-Dove Is Our Most Recent Invader

JULY 12, 1998

Yet another bird species has found South Texas to its liking. This time it is the Eurasian Collared-Dove (*Streptopelia decaocto*) that has taken up residency in our area at Six Mile, near Port Lavaca. Also called "Ring-necked Dove," due to its characteristic black collar, the name "Eurasian Collared-Dove" is preferred by the American Ornithologists' Union, the organization that establishes common and scientific names for all North and Central American bird species. Also, the closely related and very similar Ringed Turtle-Dove (*Streptopelia risoria*), which also exists in a few scattered North American cities, is a different species all together. And to make matters more confusing, the two species sometimes hybridize.

The Collared-Dove is somewhat larger than the Ringed Turtle-Dove, and the former's gray-brown color is darker than the almost whitish Turtle-Dove. But most important for identification, the calls are quite different: that of the Collared-Dove is three syllables, like "kuk-kooooo-kook," with emphasis on the second syllable; the call of the Turtle-Dove is two syllables, like "kooeek-krrroooo (aw)."

This is not the first Texas record of Eurasian Collared-Doves, as there are additional reports from High Island, Galveston, Texarkana, Austin, Johnson City, Canyon, and Farwell. And now a population at Six Mile. But the Six Mile birds appear to be the first verified record from South Texas. What's more, these birds have moved in and nested.

I received my initial information from Carl and Billie Mauer of Six Mile. They had read my "Nature Note" in the March 1, 1998, *Victoria Advocate* about the increase of White-winged Doves in the state, in which I mentioned other newly arrived dove species and expressed an interest to learn about new arrivals. They called to let me know that they indeed had some resident Ring-necks. A couple days later, I drove over to the Mauer's to see for myself, and sure enough, I saw several Ring-necked or Collared-Doves. Carl told me that his doves had recently nested in an adjacent pine tree, producing three youngsters, and the five birds come to a ground feeder every evening. I found two individuals and heard their very distinct song; almost Barred Owl-like with three deep coos.

Carl also told me that at least two additional populations were nearby, near the bay in Royal Estates and on School Road. Although I could not find any birds along School Road, I did locate the Royal Estates birds, at the home of Gloria Mills. She showed me a recent nest and one individual bird that I was able to photograph. Gloria also told me that her birds had been resident for four years and that they were present around her home year-round.

But what is most interesting about this species of dove is its spectacular spread from its original breeding grounds in India. It was introduced into Hungary in the mid-1880s and had spread westward all across Europe and to Great Britain by the 1950s. American birds likely reached eastern North America from the Bahamas, where a breeder's aviary was broken into in 1974, releasing about 50 individuals. By the late 1970s, Collared-Doves were found in South Florida, and by 1987, the breeding population in the Homestead area alone was estimated at 1,000 individuals.

Since then these doves have moved west and north; there are records from all across the southeastern United States and north to Nova Scotia, Canada. There also are reports as far to the northwest as Utah and Montana. There also is a population, probably local escapees, in central Los Angeles. Widespread distribution in Texas is just a matter of time.

It is fascinating to be able to follow a bird's "natural invasion" across the continent. Our abundant Cattle Egrets are an earlier example of such an invasion of an Old World species. These white, grassland birds appeared in Central America during the 1930s, moved north into Florida by 1948, and are now found throughout most of North America, at least in summer. A few Cattle Egrets are resident in the Golden Crescent year-round. It will be interesting to see how soon the very adaptable Collared-Dove can be found in neighborhoods across the state. The Six Mile birds, thanks to the Mauers, are one piece of the puzzle.

It also will be very interesting to assess whether or not increasing Collared-Dove populations have any influence on the long-time resident doves. It is unlikely that the increase in Collared-Doves will affect the smaller doves—Inca Dove and Common Ground-Dove—but populations of Mourning Doves and especially White-winged Doves may decline. And the increase of Collared-Doves eventually may also add another game bird to the list of huntable Texas species.

Summertime Is Lizard Time

JULY 16, 1995

Summertime is lizard time, when those cold-blooded reptiles are most active. In my yard at Mission Oaks, I have found 7 species to be reasonably common: Green Anole, Texas Spiny Lizard, Prairie Lizard, Ground Skink, and Six-lined Racerunner during the daylight hours, and Banded and Mediterranean Geckos after dark.

The most common of all these is the Green Anole (also known as Carolina Anole), a 3- to 4-inch lizard that changes from a deep green to brown or grayish color, depending upon the substrate.

The Texas Spiny Lizard is a stocky lizard with a rather scaly appearance. Most individuals are 5 to 7 inches in length, but an older lizard, that has somehow survived the predation of a number of birds, especially Roadrunners and Red-shouldered Hawks, can reach 11 inches. This is the "rusty lizard" that prefers woodpiles, trees, and rocky places but occasionally is found in the open. Males possess a narrow light blue area, without a black border, at each side of the belly.

The Ground Skink is less obvious, because it lives among the ground cover and humus, but may be the most common of all our yard lizards. It also is one of the poorly marked lizards, sporting a smooth, brown to golden back, with a slightly darker stripe down the back and very short legs. Rarely more than 5 inches in length, the Ground Skink will slither into the humus, if possible, rather than run away like most lizards.

The Prairie Lizard, the least common of the seven species, is a 5-inch lizard with a light brown stripe, edged with lighter stripes, down its back. Males possess two long, narrow light blue patches, one at each side of the belly, bordered with black. It seems to prefer sparse ground cover.

The Six-lined Racerunner is the most streamlined of our yard lizards, with a long, slender body and an extremely long tail, a total length of 6 to 9 inches. The color pattern on the back consists of six light and six dark stripes, and its underparts are plain. Males sport bluish underparts. This is an active, rather bold lizard that moves in a jerky fashion when hunting, but runs away very fast when disturbed. It prefers more open, harder terrain, where it can outrun its predators.

Finally, the geckos appear only after dark, often searching for insects that may be attracted to lights. The Banded Gecko is a little (rarely more than 3 inches in length), pinkish lizard with smooth skin and a black-and-white banded tail. It also is one of the lizards with a voice. The Mediterranean Gecko is normally 4 to 5 inches in length, with tan skin and a warty appearance. Native only to the Mediterranean Basin, it was inadvertently introduced into the United States about forty years ago and is now widespread throughout the southern states.

Most Snakes Are Good Snakes!

JULY 19, 1998

It is strange how many folks have a mind-set against snakes! They believe that all snakes are bad snakes, that there absolutely are no good snakes at all. That belief undoubtedly stems from childhood, and it is a long-term philosophy that is difficult to overcome. Too many folks automatically kill any snake they encounter, without even considering whether or not the species is harmful or how beneficial it might be.

For instance, a couple weeks ago, I encountered a neighbor, a well-meaning neighbor I am sure, along the entrance road where I live. He was stopping everyone who came along, asking for something that could be used to kill a snake that he had discovered crossing the road in front of him and was then escaping in another neighbor's yard. He was obviously excited and said that he had tried to run over the snake and now, after missing the snake, felt it was his responsibility to kill it.

A second neighbor and his young son had already stopped just ahead of me. We all climbed out of our vehicles at the same time to see what was happening. The snake—a huge Texas Rat Snake—was thirty-five feet off the road, making a fast getaway. The first neighbor was pointing at the snake in a rather animated way; he was obviously upset that he had not done his duty and killed that vile serpent.

I immediately walked over to identify the snake, but it already was obvious, even from thirty-five feet away, that the snake was not a poisonous species such as a Copperhead, which is fairly common in our neighborhood; a Coral Snake, rare but seen occasionally; or one of the other possibilities—rattlesnakes, Cottonmouth, or Massasauga, species that I had not seen in the area in ten years. No, it was a beautiful specimen of a Texas Rat Snake. Normally this snake is found only in woods and brushy areas, but the extremely dry conditions undoubtedly had forced it out of its typical habitat into our well-watered lawns. And it had happened to cross the road at the wrong time. If I had not come along when I did, I am sure that my well-meaning neighbor would have done whatever he could to dispose of one more "dangerous critter."

Just the opposite is true, however, for most of our snakes. Rat snakes feed on rodents of all types, from rats to mice, and on (sadly) some ground-nesting birds. Lizards, grasshoppers and crickets, and various other insects are also taken when necessary. Rat snakes are one of the constrictor snakes that wrap themselves around their larger prey to kill them and eventually swallow them headfirst.

An adult Texas Rat Snake (*Elaphe obsoleta lindheimeri*) is usually 42 to 72 inches in length, although Roger Conant, in *A Field Guide to Reptiles and Amphibians of Eastern and Central North America*, claims that one record Texas Rat Snake was almost 84 inches long. He describes the species as "blotched" with "brownish- or bluish-black" on a gray or yellowish ground color and "head often black." Also, some great photographs of an adult and a juvenile Texas Rat Snake exist in Alan Tennant's book, *A Field Guide to Texas Snakes.*

The mind-set against snakes is widespread in Texas and throughout most of the country. Perhaps it stems from the biblical story of Satan taking the form of a serpent to entice Eve to eat the forbidden fruit. But in today's society, when we need all the good creatures we still have to naturally take out those critters that can carry diseases or destroy our crops, killing snakes is not only ill conceived but illogical. To me, such action is a sad commentary on one's respect for nature, as well as on one's self-respect.

Bronzed Cowbirds Are with Us for the Summer

JULY 20, 1997

By late spring each year in South Texas, Bronzed Cowbirds increase in numbers and become regular visitors to our yards and feeders. But they can easily be confused with the far more abundant Brown-headed Cowbirds that have been commonplace since the first Northern Cardinals begin singing in spring.

The Bronzed Cowbird, earlier known as Red-eyed Cowbird because of their red eyes, is somewhat larger than their cousins, the Brown-headed Cowbird. Male Bronzed Cowbirds also possess a thick ruff on their nape, giving them a hunchback appearance. They lack the brownish head of the Brown-head but are overall purplish or bronze colored. Courtship is fascinating when the male Bronzed Cowbird displays for its harem. It will throw back its head and erect its large ruff, quivering its wings and bouncing up and down, all the while singing a high-pitched, creaky song as it circles its ladies.

Both of these cowbirds have the habit of laying their eggs in other birds' nests, where the foster parents will raise the larger, faster growing cowbird babies as their own. The larger cowbird young oftentimes shove the rightful nestlings out of the nest, where they usually perish, and if that doesn't occur, the cowbird nestling is so much larger and domineering that it gets the bulk of the food brought to the nest by the hard-working foster parents. This parasitic habit can produce severe declines in certain preferred host species. While Brown-headed Cowbirds parasitize such birds as gnatcatchers, vireos, and warblers, including the endangered Black-capped Vireo and Golden-cheeked Warbler, two of Texas's most unique songbirds, Bronzed Cowbirds seek out slightly larger birds, such as thrashers, tanagers, and orioles. Hooded Orioles, for example, have declined in several areas of South Texas where populations of Bronzed Cowbirds have increased in recent years.

Brown-headed Cowbirds are with us year-round, congregating in huge flocks in our pastures and fields during the winter months and pairing up in early spring. Most Bronzed Cowbirds move farther south for the winter months; the greatest numbers of Bronzed Cowbirds anywhere in the United States are recorded on the annual Kingsville Christmas Bird Count. They begin pairing in late winter or early spring, about the time that the earliest thrashers begin to sing and prepare to nest. And post-breeding Bronzed Cowbirds become more numerous and appear in our yards soon after they have mated and parasitized their neighbors.

While Brown-headed Cowbirds evolved on the Great Plains, where they fed on seeds among the herds of bison or on ticks found on those huge beasts, Bronzed Cowbirds appeared from south of the border. Invaders to Texas and Arizona only since the early 1900s, Bronzed Cowbirds now can be found throughout the southern half of Texas in summer. And like their smaller cousins, they are opportunists that can take advantage of a wide array of hosts, including such unexpected birds as Northern Mockingbirds and Mourning Doves. Only Northern Cardinals seem to be sufficiently aggressive to consistently keep these parasitic newcomers at bay.

Yardbirds Begin to Change in July

JULY 21, 1996

Returning home after a three-week trip to Utah and Oregon, I found very different bird action in my yard. Sure, all the same nesting species were still there, but there was a distinct change from when I left in mid-June. Now only the mockingbirds seemed to be in full song. These loud-mouthed songsters still claimed the best singing posts and still defended select territories. But even the cardinals, another normally vociferous yardbird, were not as vocal as when I left. And a couple of cardinal families were spending most of their time at my feeders. The grungy-looking youngsters were clothed in grayish pink, and their reddish bills and short crowns gave them away.

When walking my usual morning route, I noticed that Purple Martins and Chimney Swifts, with fewer Barn Swallows, were still present overhead, but the earlier Purple Martin ratio of adult birds had changed appreciably. Young martins were dominant, and none of the all dark males were evident. The adult males apparently have already started south to their wintering grounds in South America.

Most of the birds that are so vocal in spring are silent by July. Except for early morning singing, the majority have lost their urge to vocalize extensively. This is due, of course, to the decline of their breeding urge, as most of their youngsters are out of the nest, following their parents about, begging for handouts. But still, in the early mornings, I could hear half-hearted songs of Painted Buntings, White-eyed Vireos, Carolina Chickadees, Tufted Titmice, Carolina Wrens, Inca Doves, and Yellow-billed Cuckoos. And the Ruby-throated and Buff-bellied Hummingbirds that nest nearby and utilize my feeders are still present as well.

But as July continues, start watching for early southbound migrants. Although mid-summer seems extremely early for bird migration, a few post-nesting vagrants begin to move south very early. Two migrants that can be expected in South Texas include the Mississippi Kite and Rufous Hummingbird. While the Mississippi Kite nests reasonably close by at Lake Texana State Park, the nearest Rufous Hummingbird nesting occurs in Idaho and Oregon. The Mississippi Kites will soon appear overhead, diving and soaring in their search for flying cicadas. These large, gray, black, and silver raptors provide the watcher with a wonderful show of aerial acrobatics.

The tiny Rufous Hummingbirds will appear at our feeders and flowers, zipping about in their search for nectar and tiny insects. All of the early arrivals will be males; the females are still tending to family chores on their nesting grounds. And in spite of this hummingbird being smaller than our local breeding Ruby-throats, male Rufous Hummingbirds can be so aggressive that they dominate the feeders; only the larger Buff-bellieds can withstand that pressure. Male Rufous Hummingbirds are easily identified by their overall rufous plumage, and they also make a unique sound in flight, a surprisingly loud whistle produced by their wings.

Then, before we know it, the huge congregation of migrating Ruby-throated Hummingbirds, some from as far away as central Alberta and Nova Scotia, will be filling our yards with sparkling green lightning.

Mustang Grapes Are Ripe and Edible

JULY 23, 1995

Our native Mustang Grape vines are ripe with big, lush fruits this year. Last weekend we picked two huge buckets of these purplish black grapes that are growing on the oak and hackberry trees along the back edge of my yard. Daughter-in-law Sharon Nichols took the fresh, juicy fruits home with her, where she will make grape jelly.

The raw fruit is extremely tart, although the pulp itself is extremely sweet and tasty. They can be eaten by sucking out the fleshy meat and discarding the skin that is so caustic that it can actually cause a reaction in some folks.

Our native grapevines seem to grow best on the abundant oak mottes in our area, and they often grow huge trunks that wrap around their hosts. By mid-summer, when the grapevine foliage and fruit are at their peak, they seem to dominate many of their host plants. Some oat mottes can look like huge pyramids of grape leaves.

The Mustang Grape, known to scientists as *Vitis mustangensis*, is the most common of the fourteen native grapes in Texas. Although all of these have been used for jellies and domestic wines in the past, Mustang Grapes are the largest and most prolific. Summer grapes are common along the river bottoms.

We got to the abundant fruits just in time, because they were rapidly being eaten by several bird species. Apparently the birds discard the skins as well, because I found grape skins littering my yard. Northern Cardinals, Blue Jays, Northern Mockingbirds, Carolina Wrens, Great Crested Flycatchers, and even woodpeckers were sharing the harvest. I watched a family of cardinals pick and eat several grapes one after the other. Each would pick a grape, squeeze it in its bill to discard the skin, and then work it around in its mouth to consume the pulp. In most instances, the cardinals spit out the seed, but some individuals actually swallowed the pulp, seed and all. The seed was undoubtedly discarded at a later time.

It seems amazing how many birds and other animals, including our local Fox Squirrel, utilize various fruits that ripen in the summer. Many species, such as the Great Crested Flycatcher, prefer insects during much of the year but are able to change their diet to fruits, almost exclusively, when they become available. Many of the neotropical migrants, birds that nest in North America but spend their winters in the tropics of Mexico and Central and South America, feed almost exclusively on fruit during the winter months but prefer insects and other invertebrates in spring and summer. The American Robin is one of the best examples; it changes its diet almost overnight, once the youngsters are fledged, from worms and insects to fruit.

My grapevines will be bare of fresh fruit in a few days, but the grape season provides good jelly and lots of wildlife viewing.

Two Kinds of Treefrogs Are Local Residents

JULY 26, 1998

It is more than likely that the two species of treefrogs that are currently residing in my yard have been present for many years, but this is the first year that I have found them both. Both species reach the southeastern edge of their range in South Texas. And both seem fairly common this year, probably because my yard represents a haven for many of our water-starved creatures.

Most abundant of the two is the Gray Treefrog, about 2 inches in length and primarily gray with brown and black colors. It can look silver-gray at times, especially in the middle of the day when it is resting in a protected corner. The colors change somewhat, depending upon the activity or environment of the treefrog. Other distinguishing features include a whitish spot beneath the eye and a bright orange or golden yellow color on their concealed hind legs.

The second treefrog is the Green Treefrog that is usually bright green, although some individuals can be nearly yellow, with a long yellow to whitish patch that runs from just below the eye to the flank. Slightly larger than the Gray Treefrog, it has many of the same behavioral characteristics of hiding out during the heat of the day. This treefrog often can be seen on our house windows at night, where they come to prey on insects attracted by the indoor lights.

The Green Treefrog is also known as a "rain frog" because it tends to sing just before rainstorms. However, Green Treefrogs can be heard during almost any weather condition and almost any time of day or night. Their call "has a ringing quality, but is best expressed as 'queenk-queenk-queenk,' with a nasal inflection. It may be repeated as many as 75 times a minute," according to Roger Conant in his *A Field Guide to Reptiles and Amphibians of Eastern and Central North America*. The call of the Gray Treefrog is more of a slow musical trill.

Both of our treefrogs possess long legs and webbed toes with adhesive discs that allow them to walk about on almost any surface, even a pane of glass. They also possess a vocal sac under the throat that inflates like a round balloon when singing. Both species change colors to better match their surroundings, providing them a marvelous method of protection. And, like all of our amphibians, they enter the water to lay their eggs; the tadpoles must live in water until they reach a subadult stage and can come out onto the land. All the treefrogs are classified by herpetologists (scientists who study amphibians and reptiles) in the genus *Hyla*, that also includes the well-known Spring Peeper.

There are a few other differences between our two species of treefrogs. Habitat preference is the major one. Although both may be found around the yard, the Green Treefrog prefers wet areas, living in swamps and along borders of stream and lakes where there is plenty of dampness. And it generally stays on or near the ground. Conversely, the Gray Treefrog is an arboreal amphibian that spends much of its time in trees and on shrubs; it normally enters the water only during the breeding season.

The presence of two species of treefrogs is but one more indication of the amazing biological diversity of South Texas. And it also is a reminder that our yards can be an important refuge for a myriad of creatures if we care. Keeping our plants well watered and our environment free of pesticides can help such critters like treefrogs make it through these long, hot summer days.

The Green Treefrog is our most common treefrog.

Male Hummingbirds Love 'Em, Leave 'Em

JULY 27, 1997

It seems impossible that some of the southbound fall migrant birds are already arriving in South Texas. Only a few short weeks ago, we were watching the hordes of northbound migrants that were passing through our area en route to their nesting grounds to the north.

Barely four to six weeks later, some of those same birds are heading south to their wintering grounds in Mexico and Central America. Although the majority of the neotropical migrants are still feeding young far to the north, a few, such as male Rufous Hummingbirds and a number of shorebirds, are already back in Texas.

It has been said, and truthfully so, that male hummingbirds are little more than brightly colored "promiscuous rakes." They have little or no home life. In spring they migrate north to ancestral breeding grounds where they establish territories, which they may defend against other hummers to their death, performing fascinating aerial displays to attract the ladies, mating whenever the opportunity occurs, and heading south as soon as all the available females are on a nest.

It is the female hummingbird that selects the nest site, builds the nest, incubates the two or three eggs, and feeds the nestlings. Male hummers are long gone halfway through the nesting season. They search out lush mountain meadows with lots of flowers, where they usually remain until late summer or fall. Sometimes, male hummers passing through South Texas will find suitable conditions to keep them in one place for a few weeks or, rarely, into fall or even all winter.

Shorebirds share similar behavior. Many of the shorebirds that we observed along the South Texas coastline in March and April continued northward to northern Canada or Alaska. Although some shorebirds are like hummingbirds, with males departing their breeding grounds soon after mating, others are caring parents. But the great numbers of shorebirds that utilize the northern tundra to nest experience a very short summer with extremely long days that permit a faster nesting cycle. Many nestlings are fed twenty hours a day, and many of the young are precocial (capable of a high degree of independent activity from birth): they follow their parents about, feeding on the abundant tundra insects, within hours of hatching.

The long days and abundant food lead to a very swift completion of their nesting cycle. So, the coastal wetlands in Texas, like a narrow part of a North American hourglass, experience great numbers of southbound shorebirds by mid-summer. And wherever masses of shorebirds are found, raptors are not far behind.

Of the hundreds of post-nesting birds that can be found in South Texas in late July and early August, my personal favorite is the Rufous Hummingbird. Its arrival somehow represents a very distinct change in the characters that utilize our yards. The summer resident Ruby-throated and Black-chinned Hummingbirds are less territorial and aggressive. Only the much larger Buff-bellied Hummingbird dominates a male Rufous Hummer. But it is the Rufous Hummingbird that controls the choice patches of flowers and a favorite feeder that were earlier utilized by the resident Ruby-throats and Black-chins. Even in September, when the yard can be filled with 50 to 100 southbound Ruby-throats, the one or two Rufous hummers that have managed to stay are dominant. Rufous males are pugnacious characters with a personality all of their own!

Amphibians Are Few and Far Between This Summer

JULY 28, 1996

The drought in South Texas has affected every living thing, including our native frogs and toads. It wasn't until I almost stepped on a tiny amphibian in my yard that I realized how scarce they had been this summer.

Normally, by now, I have seen numerous Southern Leopard Frogs, Gray Treefrogs, and Gulf Coast Toads, and at least a few Squirrel Treefrogs and Eastern Narrow-mouthed Toads. But my recent discovery, actually a new species for my yard, was my first amphibian sighting of the summer.

I naturally captured it, put it in a jar, and brought it into the house for a closer look. It was a tiny toad, only about 2 inches long and with few noticeable features. It did have one oval parotoid gland behind each eye, the surefire way to identify a toad from a frog. It also was covered with tiny warts, most of them reddish or greenish on top. But what species was it?

I then hauled out my Conant, *A Field Guide to Reptiles and Amphibians of Eastern and Central North America*, and began to check the toad illustrations, and sure enough, plate 44 included my little plump toad: Texas Toad. The text mentions that a key identifying feature is the underside of the foot, which has two tubercles, with the inner one being sickle shaped. This required a hand lens, and before long I was examining my toad's foot close-up. There they were, two tubercles, including the inner sickle-shaped tubercle. No doubt, now, of its identity. My yard herp list was now 24, 7 amphibians and 17 reptiles (9 snakes and 8 lizards).

Further reading revealed that the Texas Toad sings a "continuous series of loud, explosive trills, each ½ second or more in length. Like a high-pitched riveting machine. Trill rate 39 to 57 per second." Conant also mentions that the Texas Toad breeds from April to September, after rains.

The previous night we had had a hard, fast rain that must have brought out the Texas Toads. And so, after dark I stood outside on my deck, listening for the "high-pitched riveting machine" songs of my new acquaintance, and sure enough, at least two songs echoed across the backyard. Texas Toads were singing their unique love songs to any female Texas Toads within hearing distance. I wished them well.

August

Daddy Longlegs Are Common in Summer

AUGUST 4, 1996

Daddy longlegs, also known as harvestmen, have been out and about since early spring, but in summer the young daddy longlegs that emerge from overwintering eggs hide under boards, rocks, and other objects until they are fully grown. Not until then do they usually venture out into the open, appearing in our fields and gardens and in our barns and other structures, and we become aware of their presence. In fact, the "harvestman" name comes from their usual appearance during the summer harvest. In Europe, large numbers of Harvestmen are considered signs of a good harvest, and it is unlucky to kill one.

Most often, daddy longlegs are lumped with spiders, but they actually belong to a unique family of arthropods known as Phalangiidae. They differ from spiders in that they have no constriction, or waist, between their front part and abdomen. Also, their legs are much longer than that of spiders. This gives them a rather awkward appearance, but they are able to move surprisingly fast when necessary. And they also have a strange habit of moving up and down, like doing deep knee bends, when disturbed. Finding congregations of several dozen daddy longlegs gyrating in unison can be unnerving.

Their eight legs seem extremely fragile, and they sometimes get entangled in cracks, weeds, or whatever, but when that occurs, they simply discard the entangled leg and move on; they are able to grow a new one in no time. Their legs, however, are stronger than they appear.

Another major difference between daddy longlegs and spiders is the former's lack of silk glands. They, therefore, are unable to spin webs. And they lack poison glands of any kind; they defend themselves by emitting a foul odor. Daddy longlegs feed on spiders, mites, and small insects, which they run down and capture. They also suck juices from soft fruit, vegetables, and decaying material.

About 200 kinds of daddy longlegs occur in North America, some with a 3-inch leg span, but they all look basically alike. The females lay eggs in the ground, under rocks or in crevices in wood prior to the first frost, and the majority do not survive the winter. Here in South Texas, some individuals make it through the winter by hibernating under rubbish and in other damp, warm locations. The survivors appear in spring, but they do not become commonplace until the new crop is out in summer.

Frigatebirds Are Nature's Pirates on the Gulf Coast

AUGUST 6, 1995

I recently discovered five magnificent frigatebirds circling high over Port O'Connor. They flew so high that a casual observer might have ignored them, thinking they were only common Turkey Vultures, which also are large and dark. But a second look through binoculars revealed very different birds, even larger and more graceful in flight than Turkey Vultures.

A frigatebird's wingspan measures 90 or more inches, compared to fewer than 70 inches for a Turkey Vulture's wingspan. And their flight pattern is also altogether different: frigatebirds possess narrower wings, a longer bill, and a longer and deeply forked tail, reminiscent of a pair of curved scissor blades. Vultures possess broader wings and a squared tail, hold their wings in a shallow V, and rock slightly from side to side in flight.

Male frigatebirds are glossy black with metallic purple and glossy green on their back and wings and a blood red throat, or gular sac, that inflates like a huge balloon during courtship. Adult females are brownish with a white chest. Juveniles possess a white chest and throat.

These huge creatures come north into coastal South Texas after nesting in isolated mangroves along the Gulf Coast from Veracruz southward. Some arrive in our area as early as mid-May, but they are more likely to be seen along the coast or over the Gulf of Mexico from July to mid-September. And in spite of their size, they are not often seen. They usually soar very high overhead, riding the thermals and upper winds; they even sleep on the wing.

Frigatebirds are more likely to be seen out over the Gulf waters, where they prey on squid, small sea turtles, jellyfish, and various other food. Feeding in flight, they capture their prey above or just below the surface with their long, serrated, hooked bill. They also parasitize other seabirds by chasing down smaller birds with their catches. Watching one of these huge, long-winged frigatebirds can provide a thrill; they are able to keep up with and overtake most smaller species. They eventually harass the other bird into dropping its catch, which the frigatebird either catches in midair or scoops off the water. Their piracy has provided them with two common names: "frigatebird" and "man-o'-war bird." And they also are known as "hurricane birds" because of their habit of appearing over land just prior to storms at sea.

Frigatebirds are the most aerial of all the seabirds, and except when nesting, they spend their entire lives at sea. They never swim or rest on the ocean surface. Their long, narrow, pointed wings have a greater surface area in relation to their body weight than those of any other bird, and the skeleton of a frigatebird weighs only four ounces, less than the weight of its feathers. Therefore, frigatebirds are able to soar with a minimal amount of uplift, even at about 50 feet above the ocean's surface. But sightings have been reported at more than 4,000 feet elevation. Their presence along our coastline provides the opportunity to watch one of nature's true masters of flight.

A Second Brood of Carolina Wrens

AUGUST 10, 1997

A second brood of Carolina Wrens are out and about, following their parents around my yard. The little, short-tailed fledglings, still with a few fluffy feathers sticking out of their buff-colored plumage, must not have been out of the nest for more than a few days, but they already know where and how to find food.

These wrens seem to search every possible nook and cranny in my yard, from the furrowed bark of the Live Oaks to the lattice around my back deck. And I noticed that one youngster seemed curious enough to check out the seed feeders. Although Carolina Wrens rarely utilize seeds, they certainly are able to find a number of insects about the feeders.

The most recent nest was located in the neck of my fifth wheel trailer that was sitting out in the yard. It must have been good and warm at mid-day. But over the years, I have found Carolina Wren nests in an abundance of other places, including flowerpots containing plants out in the yard as well as empty ones in the garage, empty buckets, piles of brush, and a number of natural cavities. I once found a nest started in the pocket of a pair of field pants hung on the line.

I suppose, however, that I am attracted to Carolina Wrens more because of their wonderful and varied songs than their curious behavior. Carolina Wrens possess an amazing repertoire of songs that vary from the typical "tea-kettle tea-kettle tea-kettle" to "wheedle wheedle wheedle." Other common songs include "cheerry," "which jailer," "ught-ley," and "come-on," almost always in a series. Each male is known to sing from 27 to 41 different song types, singing one repeatedly before switching to a different song type; neighboring males frequently match song types. And pairs often duet.

Carolina Wrens also possess an array of calls that range from scolding and metallic rattles to clinking, musical trills and toadlike squeaks. At times, the adults may continue singing for ten to fifteen minutes, almost nonstop. Like other songbirds, they are most vocal in the mornings and evenings, but Carolina Wrens, like our common Northern Mockingbirds, can often be heard all day long. And they are most active during courtship.

The Carolina Wren is one of our most common resident birds. Like cardinals, they are year-round residents. And they are our largest wren, about 6 inches in length. They possess a rust to buff plumage, with fine black streaks across their wings and short tail, and bold white eyebrows, one of their most distinguishing features. Although Carolina Wrens are common throughout the eastern half of Texas, from Brownsville to Texarkana and west to San Angelo and Abilene, they are not found in the far west or to the north. They are truly wrens of the eastern and southern portions of North America and south into eastern Mexico. And in my yard, they are most welcome!

Are You Ready for the Annual Hummingbird Invasion?

AUGUST 11, 1996

Very soon now South Texas will experience one of Mother Nature's most exciting events, millions of Ruby-throated Hummingbirds pouring through our area en route to their winter homes in the Tropics. My yard can literally be filled with these tiny, fast-flying gems of the bird world. As many as 30 to 70 individuals (it may be twice that number) feed at eight to ten hummingbird feeders from mid-August to late September, peaking around September 5–15.

Ruby-throated Hummingbirds can be superabundant in the fall.

Rockport's annual Hummer/Bird Celebration celebrates this spectacular flight of hummingbirds. In 1995, more than 2,500 people registered. It is quite an occasion, with field trips, talks, banding demonstrations, and up to one hundred booths of bird- and nature-related items for sale. Great fun!

But are you ready for the hummers at home? Preparation requires little more than a hummingbird feeder, available at dozens of stores, filled with sugar water and placed in a shady spot in the yard. However, the proper use of hummingbird feeders requires some attention. First and foremost, the feeder should not be left out in our hot weather for more than three to four days. So fill the feeder only enough to last for four days, and then bring it in and clean it thoroughly before refilling it and placing it back outside. If you get mobbed by the waiting hummers, you may need to use two or more feeders, cleaning one or two daily.

Hummingbird water can be purchased or prepared at home. The commercial food is safe and may stay fresh longer than plain sugar water. I use plain old well water mixed with cane sugar at a ratio of 1 part sugar to 6 to 10 parts water; what is left is stored in the refrigerator. Red food coloring is not recommended; it may harm the hummers. If hummers are present but do not come to your feeder, hang strips of red cloth from the feeder as an additional attractant.

And what about ants and other insects? Ants can be a real nuisance, but these insects can be controlled by using an ant guard (built in on some feeders) or placing some Vaseline on the hanger far enough away from the feeder so that the hummer does not get any on its feathers. Wasps can also be a problem, especially if the sugar water is spilled, but most feeders are built so that these insects cannot reach the sugar water inside.

Plantings can also attract hummingbirds, besides being wonderful additions to the yard. My favorite include the various pentas, lantanas, and honeysuckles, as well Firecracker Bush, Firebush, Cigar Plant, and Turk's Cap. The abundant, native Tropical Sage, with its bright red flowers, which I try to save on each lawn-mowing, is also extremely popular. Other excellent hummingbird plants include Cardinal Flower, Cherry Sage, columbines, Crossvine, Desert Willow, Red Yucca, Shrimp Plant, Trumpet Vine, and Tree Tobacco.

Enjoy your hummers!

Nature's Cycles Are Not Always Apparent

AUGUST 13, 1995

A pair of Cottontails has been using my backyard for several months, feeding on fresh grass and herbs and frolicking about like puppies. They seem so trusting when I approach them, running away only halfheartedly and then stopping to look me over before eventually heading for cover. At other times one allowed me to approach within a few feet. They have given me a considerable amount of pleasure.

My pair of Cottontails was recently reduced to one individual by a pair of Gray Foxes that have also begun to frequent my backyard. Then last week I found the remains of a second freshly killed Cottontail behind the shed; one of the Gray Foxes ran off as I approached.

My initial reaction was anger, as I have thoroughly enjoyed the Cottontails. But on the other hand, Gray Foxes also have their place in the outdoors, and eating Cottontails is truly part of their nature. In fact, the foxes would not have been there if it were not for available food. And I suppose that they will move on to another area now they have depleted their immediate food supply.

I experienced a similar situation two years ago with roadrunners. After an absence of three years, they (two adult roadrunners and eventually a family of four) suddenly appeared and remained nearby for more than a year. My yard was practically overrun with lizards, especially anoles and Scaly Lizards, when the roadrunners first appeared. The roadrunners took full advantage of that food supply, actually catching lizards right next to my deck. My lizard population noticeably declined within a few months, but once the roadrunners moved on to more productive feeding grounds, my lizard population rebounded. From all appearances, the roadrunners should be back in my yard any day now.

Cycles in nature actually occur all around us, even though we don't always recognize the signs. The cyclical nature of all wild things has long been an important ecological principle. There are no averages in nature!

I recall college classes in wildlife management, where we were told about synchronous ten-year cycles in populations of lynx and Snowshoe Hares in the far north. Several Arctic predators, such as Snowy Owls and jaegers, are closely in sync with lemming (Arctic rodent) populations, and birders in the lower states find more Snowy Owls following a year when there has been a lemming die-off. And who hasn't heard of the relationship between wolves and moose at Isle Royale National Park (in northwest Michigan) and elsewhere?

It is so easy to view nature as a snapshot in time, but it is necessary to perceive Mother Nature in a longer time frame if one is to understand any part of her. For many of us, that will require a lifetime.

Parasitic Wasps Are Commonplace

AUGUST 16, 1998

I am amazed at the abundance and variety of parasitic wasps that are present in summer. Many of these are tiny creatures, some even the size of or smaller than an aphid. I have encountered most of the smaller species during my search for butterflies. Using a pair of binoculars that can focus to about 4 feet has opened an entirely new world of miniature creatures that I previously had ignored. But now, along with being able to identify even the tiny Skippers (butterflies), I have also discovered many other tiny insects. And the parasitic wasps are some of the most interesting.

The largest and most obvious of these wasps are the various spider- and cicada-killers. The best known are the Pepsis Wasps, blue-winged creatures with a red body that can reach 2 inches in length. The largest one, although rare along the Gulf Coast but far more abundant throughout the southwestern deserts, is known as the "tarantula hawk." A slightly smaller species—*Pepsis elegans*—with orange wings and a blue body, is far more common in the Golden Crescent.

These impressive wasps spend a good deal of their time on the ground walking about in search of spiders. Once an appropriate spider is found, it will sting its prey, depositing just enough venom to paralyze but not kill it. When the spider is subdued, the wasp will then carry or drag its prey, depending upon the size of the spider, to a hole that it already has excavated in the ground. It will then cram the spider into the hole, lay eggs in the spider's body, and cover over the hole with the excavated debris. The wasp eggs will hatch within a few days, and the larvae will consume the body of the living spider.

Another parasitic wasp is the Cicada-killer or *Sphecius speciosus*, which looks all the world like a huge Yellow Jacket; it utilizes a very similar process. This wasp, however, finds its prey above ground in trees. Once it discovers an appropriate prey, it will jab its stinger into the insect's nerve center, paralyzing it. When that occurs, the wasp and prey usually fall to the ground. She will then drag the cicada back up the tree to where she can carry her prey in a direct glide to her burrow. The size of the cicada negates any chance of carrying such a load up and over the vegetation. It is much easier, apparently, to drag her prey up the tree and then glide it to its burrow than by dragging her prey through the grass and other vegetation. Unlike the Pepsis Wasp, which lays numerous eggs in the body of a spider, the Cicada-killer deposits a single egg on the body of the paralyzed victim. And since its burrow usually contains several chambers, it takes several cicadas before its chores are complete.

None of these adult wasps actually feed on their prey. Instead they utilize sap or nectar from flowers. The wasp larvae, however, remain underground, feeding on their victim for several days until they reach full size and their food is depleted. Then, rather than emerging as adults, they spin a silken cocoon in which the larvae remain and develop until the following season. Finally, in summer of the following year, they emerge as adult wasps. And the process begins all over again.

What Do Mockingbirds and Reddish Egrets Have in Common?

AUGUST 17, 1997

Watching a Northern Mockingbird find insects the other day reminded me of the way Reddish Egrets find their prey in a totally different environment. While mockingbirds utilize our yards, fields, and open woodlands, Reddish

Egrets live along the coast and rarely occur inland. Yet both of these very different birds utilize the same basic method of finding prey. Mockingbirds spread their wings and tail, so their white wing- and tail-patches show well, as they move about on the ground or in low shrubbery; the contrasting movement startles insects and other small prey. As soon as they move, the mockingbird quickly captures its prey.

In the case of the Reddish Egret, it spreads its large wings, like an umbrella, and wades about in the shallow waters. Its movement in the water and its spread wings startle prey species, similar to the mockingbird. It also provides shade so that the egret can see better, without the glare of the sun off the water. Reddish Egrets hunt this way for long periods of time, running down fish and other prey species and capturing them with swift strikes of their powerful bill.

Similar methods are utilized by a wide variety of birds throughout the world. Even some of the warblers utilize these techniques. The Hooded Warbler that nests in our eastern forests from East Texas to the Great Lakes is well known for its tail-flicking behavior as it searches for insects on the forest floor. American Redstarts utilize similar behavior, flashing their contrasting wings and tail in the foliage of trees and shrubs. And in the Tropics, the Fan-tailed Warbler is famous for its tail-wagging and tail-swinging behavior.

But on inspection closer to home, one can find an additional variety of hunting behavior in other birds. For instance, if you have an opportunity to observe a group of wading birds fishing in a shallow pool, you can see that each species utilizes a slightly different technique. The common Great Egret, a tall, all white bird with a large yellow bill, stands very still, usually on the pond's edge, waiting for passing prey. The Little Blue Heron will often wade about, attempting to disturb prey. The smaller Snowy Egret, all white with a black bill, black legs, and yellow feet, will sometimes fly low over the water with its yellow feet dangling, like a lure, and grab whatever the movement attracts.

The much larger Wood Storks that visit our ponds in summer are touch-feeders. They probe underwater for their food. Their delicate mouth parts detect fish and other prey, which they grab without ever seeing into the water. Most herons and egrets are rarely as active as the touch-feeders; they must see their prey first.

Tricolored Herons, a medium-sized heron with a white belly, bluish neck and wings, and a yellowish bill with a blackish tip, have developed a combined hunting technique. These waders can be inactive, like the Great Egret and Little Blue and Great Blue Herons, or they can be fairly active. The active ones walk about a pond or drag one foot to stir up the bottom sediments to frighten prey; then they will dash after the fish or whatever prey species that makes a run for it.

And then there is the smallest of our waders, the Green Heron. It will stand on the shore or on a branch low over the water, waiting for passing prey, or it may stalk its prey. It can take a position in which seems impossible to maintain its balance. If waiting or stalking doesn't work, it may actually use "bait," dropping a twig or piece of bark to lure a fish within striking distance. For Green Herons, that method seems to work well enough, as plenty of Green Herons also are about in summer.

Plant Galls Are Most Intriguing

AUGUST 18, 1996

The feeling of mid-summer is everywhere, from the fields and pastures to the oak mottes, river bottoms, and drier pine stands. Each community has its own distinct personality, its own assemblages of plants and animals. Each is an intricate web of nature that contains marvelous inhabitants with even more fascinating relationships. The ecology of plant galls is one of the most intriguing.

Although we don't know much about the physiology of galls, we do know that they are begun by insects. We know that galls are the plant's response to an insect that lays its eggs either on the host plant or inserted in the tissue, and when the larva emerges, it makes its way to the plant's meristematic tissues that are capable of dividing and multiplying. Once the larva reaches these tissues, it secretes a substance that stimulates the cells, causing them to grow abnormally, producing a gall. The larva then feeds on the growing tissues, develops into an adult, eventually emerges, and departs its host undamaged.

We see galls on plants almost daily when we are out and about, although we may not always recognize them for what they are. Oak "apples," round and brown and sometimes an inch or two across, are best known. But galls come in numerous sizes and shapes and can appear on hundreds of plant species. For instance, galls can occur on a variety of leaves, while others form on twigs and branches, roots, blossoms, fruit, and even seeds. Galls are especially numerous on oaks, willows, roses, legumes, and composites.

If you take a sharp knife and cut into a growing oak gall, you are likely to find a juicy, spongy white substance and a central larval cell, holding a small grub or young gall wasp. You might also find a number of other creatures.

What is most fascinating about galls is that the insect that emerges may not be the responsible agent. It might be a "guest" insect. Many galls are complicated communities. For instance, the pinecone gall is a veritable insect apartment house. This particular gall is known to house its maker, as well as up to 30 more species, representing guests, parasites, and transients. Yet the remarkable thing is that those made by the same species of insects always have the same form and always occur on the same species of plant.

The insect world is so complicated that it boggles the mind. Yet it is a tiny fraction of our existence. Can anyone imagine the intricacy of all the rest of our world?

The Fascinating Black-and-Yellow Garden Spider

AUGUST 20, 1995

There is hardly a yard anywhere in South Texas in summer that does not possess one or several Garden Spiders, often called the Black-and-yellow Garden Spider; scientists have named it *Argiope aurantia*.

Among the largest and showiest of spiders, the adult Back-and-yellow Garden Spider possesses an inch-long black abdomen with yellow or orange markings. The front part of its body is generally gray above and yellow below, and its eight legs are long and velvety. But this spider is most notable because of the huge web that the female spins between all types of structures, from trees and branches to sheds and barbecue pits.

Garden Spider webs, usually 3 or 4 feet in diameter but occasionally up to 8 feet across, are surprisingly strong and flexible. It is said that spider silk is the strongest natural fiber known, that even steel drawn out to the same diameter is not as strong. The strong silk threads from Garden Spiders were once used as crosshairs in telescopes and other fine optical instruments. I can well attest to its strength; mowing or watering the lawn, I will occasionally crash into one of these webs, and it takes several tries to remove the webbing from my arms or face. A damaged web will usually be reconstructed during the evening hours.

Each Garden Spider web has a distinct zigzag band of white, sticky silk running vertically through the center. This white band may also help birds see the net so they do not fly into it. The Garden Spider, unlike many other spiders, does not have a nest but remains either in the center of the web, hanging head down or hiding nearby. The movement of key strands of the web signal whenever prey becomes trapped and attempts to escape. The strength of the webbing suggests that it can capture and eat rather large prey, from a wide variety of insects to small lizards and even hummingbirds. From observing the various webs in my yard, flies, small and large, are the spider's number one prey species, but I have also found wasps, grasshoppers, a dragonfly, and even a gecko lizard entangled.

Female Garden Spiders are considerably larger than the males and generally command the web; males construct smaller, less noticeable webs in less obvious locations. By fall, the females lay eggs in large pear-shaped cocoons

The common Garden Spider weaves a most distinct web.

with a brown paperlike surface, hung by threads among the trees and shrubs. The young hatch during the winter months but remain in the cocoon until spring. The adults usually die during the cooler winter months.

So goes the life and times of our lovely Garden Spider.

Leopard Frogs Are Very Adaptable Critters

AUGUST 23, 1998

Even during the last several weeks without rain, leopard frogs have flourished. They apparently are able to live in some of our driest habitats, even without the more typical wetlands that they prefer. These frogs seem able to get along just fine, so long as they can find grasses or brushy areas to provide shade from the hot and dry daytime conditions. And, except when they are disturbed by lawn-mowing, they will hold tight to these microhabitats during the day.

Leopard frogs are what almost everyone visualizes when thinking of frogs. Most individuals are 1 to 3 inches in length when squatting but may be twice as long with their legs extended. Some of the largest can be 12 inches or more, and those individuals are big enough to offer a delicious meal of frog legs. Leopard frogs are wonderful jumpers; some can jump three feet or more. The typical leopard frog can easily be identified by the leopard pattern of black blotches on a green background and a pair of whitish stripes that run down its back. Its legs are usually brown with black blotches edged with lighter borders.

The most common leopard frog in South Texas is the Southern Leopard Frog, or *Rana utricularia*, while the Rio Grande Leopard Frog (*Rana berlandiere*) barely reaches our area from the south. It is possible to find both species in the Golden Crescent, but those that inhabit my yard are all the southern form. The two species can be distinguished by the longer, pointed head, light spot on the tympanum (ear membrane), and greenish back of the southern. The Rio Grande Leopard Frog has a blunt nose, lacks the greenish back, and the spots are brown without lighter borders.

All leopard frogs prefer wetlands, including permanent and semipermanent pools, as well as flooded roadside ditches, where the female lays masses of eggs. These egg masses, which may include 1,000 eggs, are laid in shallow water. The female is usually enticed to the site and pursuaded to mate by a singing male. His love song is a series of guttural croaks and clucks, like the sound produced by rubbing a finger across an inflated balloon. And breeding can occur at any time of year, so long as hot or mild temperatures prevail.

Leopard frogs are members of the family Ranidae, or true frogs. This group includes thirteen species, all of the genus *Rana*. Bullfrogs are probably the best known of these, although the River Frog, Pig Frog, Carpenter Frog, Bronze Frog, Green Frog, and Mink Frog are closely related. The Leopard Frog group includes five species, among them the Northern Leopard Frog, found only in the northern tier of states and in Canada; Plains Leopard Frog, that occurs only within the Great Plains, including the Texas Panhandle; and the Pickerel Frog, found throughout most of the eastern deciduous forest region, including all of the Texas Pineywoods.

Toads belong to a separate family (Bufonidae) and can easily be identified by their dry, warty skin, compared to the moist, relatively smooth skin of frogs. Most toads also possess a cranial crest (ridge on the head) and parotoid glands, the round or oblong knobs located just behind or below the eyes. Frogs never have these features. And what's more, one cannot get warts from touching toads. However, secretions from the skin glands of both toads and frogs can be irritating to mucous membranes. Some folks are more allergic than others. So whenever touching these amphibians, be sure to wash your hands afterward. Until you do, keep your fingers away from your eyes and mouth.

All of these creatures are part of our natural environment. Appreciate them in their native habitats. We all are residents of the same planet.

Mississippi Kites Are Putting on Acrobatic Shows

AUGUST 24, 1997

Every year in late August and early September, one of nature's most interesting raptors—the Mississippi Kite—spends time about our towns and over our woodlands and fields. We watchers are provided with some amazing examples of aerial acrobatics as the kites circle and dive, swoop and zigzag overhead, capturing cicadas and other large flying insects in flight or taking them off high foliage. It is like attending a circus performance, free of charge and often without even having to leave our yards.

Mississippi Kites are medium-sized hawks, smaller than our common, resident Red-tailed Hawks and with a very different shape, but larger than the American Kestrels that spend their winter months in South Texas. Mississippi Kites can easily be identified by their slate gray plumage, black tail, and pointed wings. In a sense, they look like giant swallows. And with binoculars, you can usually see their ruby red eyes against a black eye ring and gray head. Juveniles are mottled brown and gray with a barred tail.

What makes this raptor so appealing is its fantastic flight while capturing its prey. Often it will dive with breathtaking speed, swoop, and tumble, often somersaulting in its aerial maneuvers. One can actually watch it capture and consume its prey in midair, unlike most other raptors that feed on a post or on the ground. While circling, it will hold its prey with one talon and eat its soft body parts, usually discarding the wings and head. Although cicadas are common in late summer and are one of the kite's most abundant prey species, it also will take a wide range of other insects and even larger prey such as lizards and small snakes. These gray raptors also are commonly found hawking over fields where cattle are grazing. The cattle scare up insects that are then taken in midair by the kites. And rarely, kites perch on fence posts and the like, from where they can rush out to pick off passing insects.

Mississippi Kites reside in Texas only during the spring and summer months, and they spend their winter months far to the south in Argentina and Paraguay. They usually arrive in the United States in early to mid-March, traveling in great flocks sometimes numbering in the 100s. Nesting occurs from Lake Texana north through the state into the Great Plains, west to eastern New Mexico, and east to the Carolinas.

Most of our spring birds pass quickly by, but the southbound migrants often linger at choice feeding sites for several days or weeks. Peak numbers occur in late August and early September, then decline until mid-October when all have moved south.

It is that time of year. So enjoy our aerial acrobats!

Praying Mantises Don't Always Pick on Animals Their Own Size

AUGUST 25, 1996

Along with bees, mosquitoes, ants, butterflies, and moths, the Praying Mantis is one of our best known insects. This probably is mostly due to its typical praying stance and its very distinct, almost grotesque, features: extremely long, snakelike body, triangular head with huge, bulging eyes, and powerful, angular forelegs, armed with strong spines and fitted for grasping prey.

Although they rarely fly, mantises are *able* to fly very well. And they seem to be attracted to lights at night, probably to search for prey that have also been attracted to the lights. They have been found many stories high on lighted buildings in our cities.

Mantises give the appearance of praying as they wait for passing prey species, which can range from insects of all kinds and sizes, including other mantises, to surprisingly large animals. There even are records of mantises capturing and eating hummingbirds. Mantises normally are very slow in their movements, but they are extremely swift in reaching out and grabbing prey species that they usually kill with a bite to the neck, severing nerves and leaving them helpless.

Another reason that Praying Mantises are so intriguing is their ability to look over their shoulders, behavior not normally associated with insects. Mantises overwinter in the egg stage. In fall, the female deposits foamy masses of up to 200 eggs on a twig; the mass quickly hardens into a waterproof, walnut-sized egg case. The eggs hatch the following April or May, and the little mantises drop down and scamper off. Those that remain nearby are often eaten by their larger siblings.

Full-grown mantises may be ½ to 6 inches in length, depending upon the species. Our native Carolina Mantis averages about 2 inches, the exotic Chinese Mantis is about 4 inches, and the pale green European Mantis about 2 inches. The most common mantis in the United States, however, is the nonnative Oriental Mantis, which can reach 6 inches.

Oriental Mantises are sometimes kept as pets, although each must be kept in a separate cage because of its habit of eating other mantisis. They can be fed insects, such as mealworms, pieces of raw beef, apple, potato, and other raw vegetables. They will soon learn to take food from your fingers, and they will eagerly sip water from a spoon. Mantises make very strange but fascinating pets!

September

Then the Rains Came and Brought Magical Fairy Rings

SEPTEMBER 1, 1996

What with the heavy rains of late, expect to find one of nature's truly marvelous spectacles right in your own yard. Fairy rings are circles of mushrooms that can appear overnight in lawns or pastures.

It was once believed that this circle of mushrooms marked the place where fairies danced at night, and so the name. Others believed that it was the location where the devil churned butter, leaving a ring of deadly mushrooms, and still others said that it was caused by the scorching breath of dragons.

In truth, fairy rings actually mark the edges of an underground network of hyphae, the lateral branches of a growing mushroom. The majority of the plant lives underground, but with the right circumstances, such as heavy summer or fall rains, the plant sends fruiting bodies above ground. The mushroom, as we know it, is only the reproductive part of the plant, the portion that will produce spores. The extensive network of filaments remain hidden below ground, living on decaying wood, rich humus, and similar dark, damp objects.

Normally, fairy rings occur in the same place year after year, growing larger with each flowering season. A fairy ring may extend up to 40 feet across in twenty years. In Great Britain, some fairy rings have been estimated at four hundred to six hundred years old. Most often the mushrooms that form the fairy rings are "fairy-ring mushrooms," a small, pale brown, toxic species, but fairy rings can be made by numerous species, including

A fairy ring of mushrooms can appear on lawns and in fields immediately after a heavy rain.

Shaggy Manes, Puffballs, Mica Caps, Inky Caps, Sulphur Mushrooms, Morels, Parasols, and many others.

Mushrooms are actually fungi, nonflowering plants that lack true leaves, stems, and roots. Nor do they possess chlorophyll, so they are unable to manufacture their own food. They reproduce by sending a "flowering head," which we know as a mushroom, above ground. The head then develops billions of tiny spores. These dustlike spores can be spread by the wind, scattered by raindrops, or attached to wildlife or humans. They actually are able to ride through the air for thousands of miles. Many will eventually land and germinate.

Are the mushrooms that form fairy rings poisonous? Most are, but others can be eaten. And some can be poisonous to one person but not to others. So, unless you know which species is which, it is safest to just admire mushrooms for what they are, one of nature's most fascinating plants.

Texas Purple Sage Is Full of Flowers

SEPTEMBER 6, 1998

The South Texas drought has suddenly broken wide open, and evidence of the weather change is all around us. Our lawns, gardens, and fields have suddenly come into bloom. Mushrooms seem to arise overnight, and long-stemmed Rain-Lilies and Copper Lilies will not be far behind. But of all the gorgeous flowering plants that are now in full bloom, few can compare with the common ceniza (or cenizo) shrubs. These brushy, silver-leafed plants have exploded with deep violet-purple flowers.

Three wild cenizas are native to the more arid areas of Texas, especially to the Big Bend country, but cultivated plants can be found in every corner of the state, as well as in many foreign countries. The most popular of the cultivated cenizas is also called "Texas Silverleaf" or "Purple Sage," but scientists know it as *Leucophyllum frutescens*. A second, smaller form—Big Bend Silverleaf or *L. minus*—is also cultivated and readily available in nurseries; this plant occasionally produces white flowers. The third ceniza is the Boquillas Silverleaf (*L. candidum*) that is native only in the lower Big Bend and adjacent Mexico. But all of these plants are generally called "ceniza," a Mexican word meaning "color of ashes" (referring to the color of their leaves). They have a wide assortment of additional names, including "Barometer Bush" and "Texas Ranger." In nurseries, cultivated silverleaf cenizas may be labeled "Green Leaf," "White Cloud," "Green Cloud," "Compacta," and "Convent." Cultivated Boquillas Silverleafs may be called "Silver Cloud," "Thunder Cloud," and "Rain Cloud."

To non-Texans, perhaps the name "Purple Sage" is best known, since the purple-flowering ceniza plants, which can be widespread on the open plains of West Texas, are what gave Zane Grey the title for his book, *Riders of the Purple Sage*. But it is only after summer rains that these plants put on their flowering displays.

The typical ceniza is a compact shrub, rarely over 8 feet tall, and some may be 5 feet wide. Stems are much branched and may possess brownish or whitish hairs. Leaves are alternate, mostly obovate to broadly spatulate, and covered with whitish branched hairs, making them soft to the touch. The flowers, which usually occur singly, are somewhat bell-shaped, measure about a half-inch across, and possess a five-toothed calyx and broad, five-lobed corolla, fairly typical of other members of the figwort family, such as penstemons.

Apparently, cenizas are not heavily utilized by wildlife, although deer and cattle will browse the fresh foliage, and sheep will utilize them during some years as well. Also, Barton Warnock, dean of Big Bend wildflowers, stated in his book, *Wildflowers of the Davis Mountains and the Marathon Basin, Texas,* that "Indians used the leaves in making a tea for the treatment of jaundice."

But what is most interesting is its amazing display of blooms that come on immediately after a heavy summer rain. It is that characteristic that appeals to gardeners worldwide—one more contribution from Texas to the world.

Thunderstorms Offer Marvelous Examples of Nature's Power

SEPTEMBER 13, 1998

We have seen some wonderful examples of storm clouds in recent days. Building storm clouds display Mother Nature's immense power and can be counted among one of the most awe-inspiring sights on earth.

Thunderstorms represent violent movements of air. They occur as a result of strong uplifting drafts that sometimes build the clouds to heights in excess of 75,000 feet. Meteorologists tell us that thunderstorms develop in three stages. First, small cumulus clouds build into larger masses of billowy, mushroom-shaped clouds called cumulonimbus, the familiar thunderheads that can be seen for more than 100 miles. Second, when the ascending air reaches a low enough temperature, precipitation occurs. Tiny water droplets are blown wildly around within the clouds until they join together to form larger droplets that are too heavy to remain in cloud form. Then gravity takes over, and the droplets begin to fall as rain, ice crystals, or snow. Huge downdrafts are created when this occurs as the falling precipitation cools the air below, producing the third stage. The entire cloud becomes a sinking mass of air and precipitation.

Lightning is an electrical charge within a thundercloud or between it and the earth. Charges between clouds or between clouds and the earth are released when electrical pressure becomes high enough. The first strokes are within a cloud; approximately 65% of all discharges occur there or between clouds. Lightning to the ground starts with a relatively thin "leader" stroke from the cloud and is followed immediately by a heavy return stroke from the ground. A single lightning strike goes back and forth from cloud to ground many times in less than a tenth of a second. A lightning discharge is incredibly powerful—up to 30 million volts at 100,000 amperes—but it is very short in duration; hence, the power of lightning has never been harnessed.

The total energy of a major thunderstorm far exceeds that of an atomic bomb. The sudden heat from lightning causes the compression of shock waves that we call thunder. The distance of these can be estimated by sight and sound. Light travels at about 186,000 miles per second, and sound travels only 1,100 feet per second, or 1 mile in a little less than five seconds. You, therefore, can judge the distance of a storm by timing how long it takes for thunder to reach you after you see the lightning flash. If you hear the thunder forty seconds after the lighting, then you are 8 miles from the storm.

The energy of one of our coastal thunderstorms is almost beyond imagination. I can think of few other experiences that demonstrates so vividly the power of Mother Nature. (Portions of the preceding were taken from

For All Seasons: A Big Bend Journal, by Roland H. Wauer. University of Texas Press, 1997).

Eastern Kingbirds Pass through in Flocks

SEPTEMBER 14, 1997

Among my favorite neotropical migrants is the black-and-white Eastern Kingbird. It looks a little like a Scissor-tailed Flycatcher but lacks the long tail and pinkish sides.

Eastern Kingbirds, however, are just as stately as Scissor-tails but with a very different dress. They possess all-white underparts, black face and cap, dark back and wings, and a black tail tipped with white; they also possess a narrow orange-red crown patch that is seldom visible. And they have a very distinct song, a stuttering "kip-kip-kipper-kipper; dzee-dzee-dzee." Eastern Kingbirds nest throughout our area in small numbers; they seem to prefer wet areas, nesting in low trees along the water's edge. I have found several nesting pairs along FM 1289 between Port O'Connor and Port Lavaca.

But what makes this bird rather special is its varied behavior at different times of the year. They arrive along the central Gulf Coast as early as late March, although they cannot be expected in numbers until mid-April. These northbound migrants often occur in flocks of several dozen to 100 or more. Most pass through our area, continuing northward to their nesting grounds that extend from Washington State eastward through central Canada into the Maritime Provinces and south to central Texas.

When nesting, they occur in solitary pairs and vigorously defend their nesting territory against all other kingbirds and much larger birds as well. They become extremely aggressive, even driving off crows and passing hawks, landing on their backs and giving them a peck or two. They also have marvelous displays in midair. Their courtship flights consist of flying up and down, zigzagging, doing backward somersaults and other aerial acrobatics, all with quivering wings and a widely spread tail. They have even learned to reject the cowbird eggs that cowbirds lay in the nests of many birds of the same size as kingbirds. There are very few records of cowbird parasitism for Eastern Kingbirds.

Like all flycatchers, their principal diet during their nesting season is insects. Most are captured in flight, usually by the kingbirds' flying up or out from a lookout perch; they also will hover over the ground, searching for insects, and then pounce on unsuspecting prey. But just as soon as their nesting season terminates, they revert to their winter pattern of flocking with other Eastern Kingbirds. And they also begin to feed on fruit that is usually abundant in late summer and fall.

By early September, we find most Eastern Kingbirds in flocks, flying southward toward their wintering grounds. They will often rest on wires along the highways or bare treetops at the edge of our pastures. But they gradually drift southward, so that by early October, they have pretty well left our area, although occasional flocks can be found to mid-October.

On their wintering grounds in South America, they gather in even larger flocks, sometimes numbering in the hundreds. And oddly enough, they spend much of their time foraging for berries in the tropical forests.

The Song of Our Field Crickets

SEPTEMBER 15, 1996

Field Crickets have invaded our towns, homesites, and businesses. Everywhere you look are black Field Crickets, scurrying here and there trying to find hiding places.

Normally Field Crickets are found only in our fields and woodlots and are primarily nocturnal in character; the recent rains have driven them out of their preferred habitats into conflict with people. Millions are zapped with insecticides. But they will keep coming until the weather changes, and those that are left will then go about their business as usual.

Field Crickets usually are welcome neighbors, so long as they stay outdoors. Many people consider crickets symbols of good luck. Jiminy Cricket, of Pinocchio fame, also helped establish an image. And crickets are prized for their singing and sometimes even kept in cages in people's homes.

In China, crickets were also kept for their fighting ability; cricket fights were as popular as horse races. The Chinese actually fed their crickets special diets, including mosquitoes fed on trainers' arms, and weighed them in order to classify them for fighting.

Many of us enjoy their cheerful evening songs, and as the nights grow longer and cooler, their nocturnal serenades increase in intensity. Before winter they must mate with a female to perpetuate their species. But only the males sing. They have three basic sound signals: a calling note, an aggressive chirp, and a courtship song to attract a female.

Singing is done with the edge of one wing rubbing against the opposite wing, creating a chirping noise. Filelike ridges, called "scrapers," near the base of the wing produce the sound. We can produce a similar sound by running a file along the edge of a tin can.

Wing covers provide an excellent sounding board, quivering when notes are made and setting the surrounding air to vibrating, thus giving rise to sound waves that can be heard for a considerable distance. The cricket's auditory organ or "ear," a small white, disklike spot, is located on the tibia of each front leg. The chirps become much higher in pitch in the presence of a female. Some of these ultrasonic sounds can reach 17,000 vibrations per second, higher than most people can distinguish. Females are easily identified by a long, spearlike ovipositor (egg-laying device) protruding from their abdomen. Eggs are laid in the ground and hatch in the spring.

Our local Field Cricket, almost an inch in length, are members of the Gryllidae family of insects, closely related to grasshoppers and mantises. They feed on a wide variety of materials, including vegetable matter, and when they get into our buildings, they can consume everything from clothing to books. However, they will not remain there and breed but will return to their preferred outdoor environment when given the chance. Outdoors they are an integral part of our South Texas wildlife.

Raptor Migration Can Be Spectacular in South Texas

SEPTEMBER 17, 1995

As the mobs of hummingbirds passing through South Texas begin to subside, another group of migrants—the much larger hawks, kites, eagles, and falcons—is increasing. Their numbers should peak in late September. On select days, up to 100,000 hawks in continuous flights of over 40 miles long have been observed in South Texas. That is something to see!

It is estimated that 95% of North America's Broad-winged Hawk population migrates southward along the Texas central Gulf Coast. Moderate numbers of Swainson's, Red-tailed, Cooper's, and Sharp-shinned Hawks, Mississippi Kites, American Kestrels, Peregrine Falcons, and smaller populations of Ferruginous, Harris's, Red-shouldered, and Zone-tailed Hawks, Bald and Golden Eagles, Merlins, and White-tailed and Swallow-tailed Kites move through our area as well. Mississippi Kites have already been evident over the treetops in area towns, where they have been feeding principally on cicadas.

But the most outstanding spectacle of the raptor migration is a circling flock of Broad-winged Hawks—especially when several hundred of these hawks begin to leave a preferred overnight roosting site at one time, usually about 8:30 A.M., and slowly ascend by circling to a point where they are out of sight.

The Broad-winged Hawk is a fairly small hawk, built very much like our common Red-tailed Hawk but with a banded rather than an all reddish tail. It is a common nester throughout the eastern deciduous forests of North America. And like many of our raptors, it is a neotropical migrant that goes south for the winter. Broad-wings spend their winter months from southern Mexico south to Peru and Brazil.

Hawk migration occurs in many parts of the world, and organized hawk watches at a few key sites have provided some amazing statistics. The best known historic sites include Pennsylvania's Hawk Mountain and New Jersey's Cape May Point, but in recent years, Texas sites have produced even greater numbers. The single most productive one is Hazel Bazemore County Park near Corpus Christi, where over 1 million hawks are known to cross over each year from late September to early October. Hawk watchers at Hazel Bazemore, a geographical chokepoint, have recorded up to 100,000 individual hawks in a single day.

Other organized Texas hawk watches occur at Smith's Point, near Anahuac National Wildlife Refuge; Bentsen-Rio Grande State Park (best in spring); Santa Ana National Wildlife Refuge; Padre Island, especially for Peregrines;

Dangerfield State Park near Longview; Devil's Backbone near Wimberley; and Falfurrias.

Hawk watching can be a most exciting outdoor activity, and one in which anyone, birder or not, can participate.

Green Anoles Are Area's Chameleons

SEPTEMBER 18, 1994

We finally arrived home from our summer-long travels, glad to be back to the daily routine and the wildlife that live in the neighborhood. Although we had a marvelous summer, full of incredible scenery and adventures, from southern California to Alaska's Denali National Park, I missed some of our common yard critters. One of those was our friendly, little Green Anole or chameleon. No other lizard seems to possess the personality of *Anolis carolinensis*.

All my other yard lizards—Texas Spiny Lizard, Six-lined Racerunner, and Ground Skink—run away and hide when approached, but the anoles seem almost to appreciate my curiosity. They sit very still watching my movement, allowing me to approach to within a foot or two before moving off only a short distance. And once settled down again, they will resume their vigil, watching for passing insects or another anole that might be intruding on their territory.

Territorial males often spread their conspicuous pinkish dewlap, or throat fan, in response to an unexpected movement. An extended dewlap signals a guarded territory. This behavior is usually accompanied by a few bobs of the head or push-ups. If a rival male continues to trespass, a heated battle, with much biting, is possible. The resident male almost always is the victor, and the intruder flees.

The anole's dewlap contains a flexible rod of cartilage that is attached near the middle of the throat in such a way that it can be thrust forward by the attached muscles. When the fan is extended, the scales become widely separated, and a bright color flashes into view.

Our Green Anoles are not always green. Their ability to change color has tagged them with the name "chameleon," after the Old World chameleon. This color change, resulting from changes in temperature, humidity, emotion, or exercise, can easily be seen as the anole moves from the shade into direct sunlight or from a dark to light object. Color changes can be striking and range from deep green to dark brown or light tan to blackish.

The color change is due to the rearrangement of pigment cells in the anole's skin. When the cells expand, they partly cover other cells and produce the brown coloration, but when the black cells constrict to tiny dots, light is reflected from the other cells, giving it a green color. Cell movement is triggered by a tiny gland at the base of the brain that produces a hormone that controls the movements of the black cells. If the gland is removed, the animal remains pale green.

Such color changes are extremely beneficial to these creatures, providing wonderful camouflage when hiding or stalking prey. Passing insects assume the Green Anole is little more than a green leaf.

The largest genus of reptiles in the Americas, *Anolis* contains more than 300 kinds of anoles. Our Green Anole is the only native North American anole besides the Key West Anole of South Florida. The rest occur in Mexico, Central and South America, and the West Indies.

The Green Anole is a slender, long-tailed lizard seldom longer than 7 inches in length. It has a long, pointed head and long legs and toes. The toes expand to form an adhesive pad for walking on slick surfaces. An anole may even walk up the surface of a glass window, where you can get a really close-up view of this amazing creature.

Blue Jays Are More Active Than Usual

SEPTEMBER 20, 1998

Numerous Blue Jays, probably two or three family groups that have flocked together since nesting, are spending an inordinate amount of time at my feeders. During the recent rainy spell, they zeroed in on my handouts, even coming to the feeders while I was in the yard. That behavior for the normally shy and elusive Blue Jays was totally unexpected.

Several of these jays still possess juvenile plumage, not the bright blue and white features so evident in the adults. Even their crest and black chest-band have not yet fully developed. The adults, however, show all the typical Blue Jay features: blue back and crest, blue wings with white wing bars, blue tail with black bands, whitish cheeks, and grayish underparts with a black chest-band.

The Blue Jay is one of our best known birds, common throughout the eastern half of North America. In Texas, Blue Jay distribution extends west through the Edwards Plateau and northwest to the eastern portion of the Panhandle. They also occur southward to beyond Corpus Christi, but they are only vagrants in the Rio Grande Valley and in far West Texas.

Blue Jays, perhaps more than most other bird species, engender mixed feelings in people. Although they are often admired for their color, togetherness, and tenacity, they often are disliked as thieves and loudmouths. Jays in general are well known as predators on other bird life, often preying upon smaller birds, their eggs, and nestlings. They also feed on a wide variety of other materials, whatever animal and plant foods that they come upon. They accept almost any handout offered from seeds to fruit to table scraps. In the wild, acorns are important in their diet, and they cache acorns in out-of-the-way places for later use. They hide these nuts in crevices above ground as well as in holes they dig in the ground. This caching behavior is common to all jays as well as to other members of the Corvidae family that includes jays, crows, ravens, magpies, and nutcrackers. The advantage of such behavior is that it allows these birds to survive even during hard times, even during periods of drought. Northern corvids are able to nest in late winter, long before the abundant seasonal foods are available.

All of the jays, of which there are 9 resident species in North America (Blue, Gray, Steller's, Brown, Mexican, and Pinyon Jay, and Western, Florida, and Island Scrub-Jays), are well-known loudmouths. In fact, the harsh "jay jay jay" calls of the Blue Jay are where their name was originally derived, but this jay also has numerous other calls. In *The Bird Life of Texas,* Harry Oberholser states that the Blue Jay's "song is a series of notes considerably keyed down to low, sweet whistles, lispings, and chipperings. Usually the bird performs this jumble while concealed in tangled vegetation or on interior branches of a tree. This song is heard infrequently from March into June." He also claims that jays possess a slurred sound like "jeer" or "peer." And, "usually in spring the bird whispers a pleasing teekle, teekle that is often joined with a whee-oodle, the latter commonly called its creaking wheelbarrow note. Members of foraging groups, particularly in autumn, converse with a chuckling kuk."

Oberholser also points out that Blue Jays often sound like other birds. The "bird's teearr cry sounds like that of the Red-tailed Hawk; the similarity of the screams is apparently more coincidental than imitative. Another call is a low throat rattle." However, whether or not the jay's abundant calls are intended to imitate other birds, they do succeed in attracting attention, and some of that attention is from some of the small birds, such as chickadees and titmice, species that are susceptible to larger predators. One can't help but speculate!

Millions of Snout Butterflies Are Now Flying

SEPTEMBER 22, 1996

Literally millions of Snout Butterflies were flying in our area during the last couple of weeks. Driving the highways in Goliad, DeWitt, and Victoria Counties, or looking outside my house at Mission Oaks, I was able to count several hundred Snout Butterflies passing by in only a few minutes.

The recent rains have produced great flights of these butterflies. And in my backyard, hundreds of Snout Butterflies were found among the hackberry trees and adjacent shrubs, flying about or perched on branches and leaves.

Many of these little butterflies, about 1 inch in length and with a wingspan of almost 2 inches, were perched with their wings open so that I was able to see the patterns on the upper side very well. Others, perched with their wings folded, were more difficult to see because their gray brown underwings are so leaflike. Those perched with open wings displayed inner orange and outer white patches on a black background, as well as their typical square-tipped wings. And on closer inspection, their long snouts were also obvious, a truly unique feature that looks like part of a leaf and is said to have evolved in this butterfly family to help camouflage them.

Snout Butterflies occur throughout Texas as well as west to southern California, east to central New England, and south into Mexico. They normally frequent woodland edges and stream courses and utilize Hackberry plants to lay their pale green eggs. The larvae, or caterpillars, which are dark green with yellow stripes, feed on the Hackberry leaves. They pupate before overwintering as adults in protected crevices.

Why are they so abundant this year? Raymond Neck, author of the 1996 book, *The Field Guide to Butterflies of Texas,* explains:

> Certain climatic conditions, such as severe drought followed by heavy rains over a large area in southern Texas, produce massive amounts of leaves on the food plants. This leaf production allows a major buildup in population numbers of larvae and, subsequently, adults. The new generation of adults emerges at a time when the foodplants have been stripped of most of their leaves. The lack of leaves and the dense concentrations of adults trigger a migration involving masses of butterflies that easily number in the hundreds of thousands and even in the millions of individuals.

Dodder Looks Like Tangled Yellow Twine

SEPTEMBER 2, 1995

Dodder is commonplace along the South Texas roadsides now, and what a strange and unusual plant! It has neither leaves or chlorophyll, but its white flowers are surprisingly attractive. They form small, dense, stemless clusters of tiny tubular and lobed structures that last only a few days. But the most noticeable part of the plant is the smooth, twining vine that forms mats on various host plants. This viney mat of fleshy stems has given it a number of descriptive names: "Angel's Hair," "Love Vine," "Strangleweed," "Tanglegut," and "Witches' Shoelaces."

Our common dodder, a member of the Morning Glory family, is parasitic and lives on only a few host species: Sumpweed, Seepwillow, and a few sunflowers. They sprout from a seed and initially develop a root system in the soil. The germinating seed produces a twining stem; this becomes parasitic by means of suckers which penetrate the bark of the host species. The stem then withers away, severing its connection to the ground, and the mature plant spends the remainder of its life living on its host. Although this process may take a year or more, the plants are most noticeable in late summer and fall. At times the yellow plants may cover rather large areas of several yards, but most individuals are less than a couple feet across.

Fred Jones lists 5 species of dodder in his book, *Flora of the Texas Coastal Bend,* and the most common dodder along our roadsides is scientifically known as *Cuscuta cuspidata*. Other local species also occur in trees. A total of 23 are listed in the *Manual of the Vascular Plants of Texas*, and about 170 species, most in the Americas, are known worldwide. The majority of species are limited to one or a few hosts and will not live on other hosts, but others will grow on a wide range of herbaceous and woody plants, even grasses.

Be on the Lookout for Bald Eagles

SEPTEMBER 27, 1998

It is again time for the annual arrival of Bald Eagles into South Texas. Southern Bald Eagles normally return to their ancestral nesting sites in September to November from their summering grounds to the north. They will remain through the winter months, nest in midwinter, and leave in March or April. Young birds may linger into early May, but all of the adult Bald Eagles and their offspring usually are long gone before the heat and humidity of summer truly set in. One rarely remains year-round.

The Bald Eagles that nest in South Texas, along the coastal plain from Nueces County to near Houston, and around lakes in northeastern Texas, are members of the southern Bald Eagle race, rather than the northern race of Bald Eagles that nest north of the state and as far away as Alaska. All Bald Eagles are long-lived and mate for life, although if one of the pair dies, the remaining bird will usually take a second mate. Adult nesters construct huge stick nests in trees usually located along the waterways. Sometimes those stick nests, which may have been used for twenty or more years, become so large that they literally can break down the tree branches. One nest was measured at 10 feet across and 20 feet deep.

Females normally lay two or three large, bluish white eggs, but more than two hatchlings is an exception. Incubation takes thirty-four to thirty-six days, and the nestlings are fed by both parents for about three months before fledging. So by the time the southern youngsters are flying, it is time for them to go north. More often than not, the adults will leave ahead of the uncertain youngsters.

Although Bald Eagles take advantage of available carrion, their diet is rather broad. Wayne and Martha McAlister write in their book, *A Naturalist's*

Guide: Aransas, that "food items found in the nests on the Aransas generally confirm the eagle's diet of fish and waterfowl: flounder, mullet, red drum; a white pelican, many American Coots, pintails, scaups, and numerous grebes; swamp rabbits and cottontails, and one armadillo that may have been picked up as carrion. One adult eagle was seen in flight carrying a struggling scaup duck in its talons. Another was observed over Dunham Bay dive-bombing an osprey in an apparent attempt to make it drop its fish."

Adult Bald Eagles are truly magnificent birds, with a snow white head and tail and a huge yellow bill that are stark contrasts to its chocolate brown body. Its general appearance as a fierce predator also is in contrast to its true character, that of a timid carrion feeder. But anyone who has watched one of these magnificent creatures for any length of time cannot help but be impressed. In fact, Congress declared the "American eagle," instead of the Wild Turkey that Benjamin Franklin preferred, as our "national bird" on June 20, 1782.

Yet in spite of the Bald Eagle being established as our national emblem, North American populations declined precipitously during the 1950s and 1960s, primarily as a result of eggshell thinning caused by pesticides and heavy metals that the birds absorbed through fish and other foods. The birds were listed as endangered by the United States and Canada in 1963, and DDT, one of the most long-lived and widespread pesticides, was banned for use in the United States and Canada in 1972. Since then, Bald Eagles have made a remarkable recovery throughout their range.

Today, one can observe one of our national birds in winter in several South Texas locations. Best bet sites include a variety of

Bald Eagles return to South Texas in late September.

fishing sites such as Coletoville Reservoir, Lake Texana, and various points along the Guadalupe and San Antonio rivers. And Dupont Victoria has constructed an observation platform along the north entrance road to the plant where an active Bald Eagle nest can be seen. What's more, the public is welcome!

The Fly-Catching Eastern Phoebe Leads the Way for Winter Birds

SEPTEMBER 29, 1996

Already some of our wintering birds are beginning to arrive in South Texas, and one of my favorites is the Eastern Phoebe.

It seems like every yard or two is claimed by one of these perky flycatchers that spends its winter days capturing insects that it finds from low perches. One will suddenly dart out and snap up a passing fly or pounce on an insect on the ground or on a tree trunk or branch. Then it will return to a favorite perch, swallow its prey, jerk its tail, and wait for the next passing tidbit.

Eastern Phoebes are one of our hardiest flycatchers; all the other members of this rather extensive family go south for the winter, but this bird migrates only short distances, remaining just south of the really cold winter weather. It is able to survive the extreme cold conditions that we experience once or a few times each winter. At such times, it is actually able to feed on a non-insect diet, consuming seeds, fruit, and even an occasional frog or small fish. Its ability to change from insects to fruit and to eat frogs and fish attests to its amazing adaptability and physical characteristics, especially a bill wide enough to capture insects in flight yet strong enough to hold and swallow small vertebrates.

The Eastern Phoebe is resident throughout the eastern half of North America, nesting just north of our area, usually on structures such as porches, bridges, and cliffs. Nests are constructed primarily of mud and moss. One New England bridge site was utilized for thirty consecutive years. John James Audubon placed thin wires on the legs of one family, said to have been the first time birds were ever banded.

Its name comes from its rather distinctive "fee-bee" calls, that it may utter singly or many in an extremely very short time. It also has a clear and sweet chip note.

This phoebe, one of three in North America (including Black and Say's), is readily identified by its size (about 7 inches), brownish back, darker head, no wing bars or eye rings, white throat, and whitish to faint yellowish underparts. It also has the typical flycatcher habitat of jerking its tail when perched, but unlike most other flycatchers, this phoebe will sweep its tail widely, down then up and often toward the side, giving it a wagging appearance.

I look forward to its return. The Eastern Phoebe is a marvelous wintertime neighbor.

October

Sandhill Cranes Are Returning to Their Wintering Grounds

OCTOBER 1, 1995

October marks that time of year when Sandhill Cranes are starting to move into their wintering grounds throughout South Texas. It is wonderful to have these large, graceful birds back in our fields and pastures. They will remain all winter, leaving for their northern nesting grounds in late April and early May. From now until May, one can find hundreds of these birds by driving the roads throughout the region.

Unlike the more famous Whooping Cranes that also winter in South Texas at Aransas National Wildlife Refuge and on adjacent coastal flats, Sandhill Cranes prefer old fields and pastures. Their needs are very different. While the Whoopers overwinter only in wet coastal marshes, feeding on fish and other marine life, as well as marsh plants, acorns, and grains, Sandhills prefer newly planted or harvested corn, sorghum, and grains. They even may range into semiarid areas west to the Big Bend country.

Our returning Sandhill Cranes often possess an odd color pattern, especially when they first arrive: their feathers are rust colored from the iron-rich feeding grounds where they spend their summer months. Normal Sandhill Crane plumage is overall gray with a whitish throat and cheeks, a bare reddish cap in adults, and dark gray legs. Whooping Crane adults are all white except for a black facial pattern and reddish cap. In flight, Whoopers display obvious black-and-white wings, while flying Sandhills are all gray, except for their whitish throat. Both fly with their heads stretched far out and legs extended; Great Blue Herons and Great Egrets, also large birds

Sandhill Cranes rank among the most stately birds.

that are sometimes confused with cranes, normally fly with their necks bent so that their head does not extend.

Sandhill Cranes may be one of our most expressive birds. They often can he heard at a considerable distance, talking to one another in their unique calls, a long, rolling, hollow rattle, like "garooooooo." Whether in flight or feeding in a field, they seem to spend a great deal of their time communicating.

Although huge flocks of 50 to a 100 or more birds can be found regularly in South Texas, these groups can usually be broken down into rather distinct family groups. Especially when they first arrive in our area, groupings of adults and one juvenile are commonplace. The young generally remain with the adults all winter and gradually, by the time they are ready to head north again, mature to the point that they are difficult to distinguish from the adults.

Normally Sandhill Cranes mate for life, and as their mating season draws close, they often can be seen dancing before their mates. Although their mating displays in Texas are less elaborate than the dancing that occurs on their breeding grounds, it is still worthy of our observations. Courtship involves loud calling and marvelous dances with head bobbing, bowing and leaping, grass tossing, and running with wings extended.

It is good to see Sandhill Cranes returning to South Texas.

Vultures Serve as Nature's Cleanup Crew

OCTOBER 2, 1995

Vultures have been called the ugliest, most disgusting creatures known to mankind. They seem always to be present at carrion, from roadside kills to predator kills and dead cattle in the pastures.

Who hasn't been forced to slow down on the roads to avoid hitting one of these large, very dark birds? Hollywood has even enhanced their reputation by using vultures to symbolize death and dying in the movies, and cartoonists often show vultures sitting on fence posts waiting for an old cowhand to "bite the dust."

Vultures are so abundant in South Texas that we often take them for granted. Both Turkey and Black Vultures can be seen soaring overhead at almost any time of the day. A short drive out of town will likely produce both species sitting along the roadside or on an adjacent post or treetop. In the early mornings, they often can be found perched with their wings spread to absorb the morning sunshine. A vulture's nighttime temperature is lower than it is during the daytime, and the spread-wing posture allows it to quickly absorb solar energy to raise its temperature to the daytime level.

As soon as the sun warms the air ever so slightly, they begin soaring in search of carrion. Turkey Vultures can soar for hours on end, hardly ever flapping their wings once airborne. They can take advantage of the slightest updrafts from winds blowing over the terrain as well as the thermals, stronger columns of warm air rising from the lowlands. The heavier-bodied Black Vultures require greater lift and use faster wing beats to get airborne and between long glides. Turkey Vultures have longer, bicolored wings, fly with their wings held in a shallow V pattern, and rock slightly back and forth; they also possess a bare red head. Black Vultures are all black and fly with their wider wings held straight out; they possess a bare black head. Their nonfeathered heads are an adaptation for feeding on the gore of carrion.

There has long been an argument about whether Turkey Vultures find their food by sight or smell. It now appears that they find food primarily by a keen sense of smell. To prove the importance of their olfactory senses, researchers placed chicken carcasses of known age in hidden locations. Those vultures almost always found the carcasses on the second or third day when they were beginning to ripen but rarely visited them when they were four days old in a state of full-blown putrefaction. They rarely fed on fresh carcasses on the first day, in spite of eating freshly killed chicken in captivity. The researchers believe that the one-day-old carcasses are too fresh to give off enough odor to be easily detected. By the second and third days, however, enough decay occurs to make the carcasses noticeably pungent. By the fourth day, the quality of the meat is severely compromised by the buildup of microbial toxins and is no longer attractive.

It appears that our abundant Turkey Vultures possess an ability to find their food by smell alone.

White-Tailed Hawk, One of Texas's Special Raptors

OCTOBER 4, 1998

Driving south on US Highway 77 and SH 35 en route to Rockport and the Hummer/Bird Celebration two weeks ago, I was treated to a dozen or more White-tailed Hawks perched on adjacent utility poles. Many seemed especially poised just for me. Perhaps it was the foggy morning, and they had not yet fully awakened and gone off to search for breakfast. But whatever the reason, I could not help but admire those stately birds—not just because of their bright, contrasting markings, but also because the White-tailed Hawk is one of Texas's specialty birds, occurring nowhere else in the United States except in Texas. Nowhere are they more numerous and easy to find than along the open highways in the Golden Crescent.

The total range of the White-tailed Hawk extends far south of our Mexican border to Argentina, but their U.S. range is limited to the coastal prairies of Texas. Vagrants occasionally are found as far west as Central Texas and the Big Bend country, and at one time, they were residents of the yucca grasslands west into Arizona. Those birds were decimated after cattle were introduced and the landscape changed from native prairies into brushlands. Nests are usually built in isolated low shrubs or trees. Although there is some northward movement after nesting, the birds are generally full-time residents wherever they occur.

Slightly smaller than our common Red-tailed Hawk, adult White-tails possess a snow white tail (with a black terminal band) and underparts, gray back and head, and rusty shoulders. The bright white tail shows up best in flight. Immature birds, however, look very different, which can be very confusing. They usually are overall blackish with a creamy breast patch. They also show a whitish patch on each side of the head.

White-tailed Hawks seem to remain perched on favorite overnight roosting sites longer each morning than most other hawks, at least during the nonbreeding season when they are not feeding young. Prey consists of almost anything available on the coastal prairie from small mammals, birds, reptiles, especially snakes, and even insects. Like other raptors that hunt the prairies, they are readily attracted to wildfires. I have seen congregations of 50 or more individuals soaring over prairie fires, stooping down to capture prey escaping from these burns. Sometimes the hawks even land and run down prey. They seem to get along with prairie fires very well.

Normal hunting activity is accomplished by either searching for prey from a perch or soaring or hovering overhead and diving on prey species. Once fed, they will begin to soar and, within a few minutes, will find an adequate thermal and ride that upward and out of sight. A soaring adult White-tailed Hawk is truly a thing of beauty!

Leafcutter Ants Are Part of a Unique Society

OCTOBER 9, 1994

Texas Leafcutter Ants are especially active in spring and fall, times when new, fresh leaves or fruits are available before colder weather begins.

In my yard, thousands of Leafcutters recently were busy transporting their green baggage from a huge Cedar Elm tree to four separate nest entrances 40 to 50 feet away. Their routes were littered with the green fruit, and the workers, carrying their loads high over their bodies, looked like miniature sailboats moving in one direction along green rivers.

The Leafcutters don't eat the vegetable matter itself but feed on tiny fungi that they "grow" within their underground chambers with pulp de-

Leafcutter Ants grow their own food.

rived from the chewed vegetable matter. Once the greenery is chewed into a pulp, the ants defecate a fecal droplet of liquid on it before placing it on the living garden. These ants and their fungi are totally dependent upon one another; the fungus is never found outside the Leafcutter's subterranean gardens.

The Texas Leafcutters are members of a large group of tropical ants of the genus *Atta* that includes about 200 species and reaches the northern edge of their range along the Texas Gulf Coast. One isolated Leafcutter species occurs as far north as the New Jersey Pine Barrens. Colonies of *Attas*, also known as Leafcutter, Parasol, or Fungus-growing ants, may contain over 5 million individuals that include one queen and thousands of workers and soldiers.

Atta workers vary in size, depending upon their responsibility within the ant community. The largest workers defend the nest, medium-sized workers forage for plant materials, and the smallest ones take care of the eggs and larvae. Soldiers are larger and equipped with formidable pincer jaws; they guard the workers from predators such as wasps, which capture soldiers in order to lay eggs in the ants' bodies. The roving soldiers follow the workers and greet one another with an exchange of chemicals and by stroking their antennae. The well-protected queen, largest of all, remains in a special chamber where her function is to produce eggs for the thriving colony. When a move is required, an egg will be manipulated to produce a new queen that will then leave the nest to found a new colony. She will carry some of the precious fungus, in a special cavity in her mouth, to the new site.

Studies of Leafcutter Ants in Panama and Costa Rica revealed that different colonies have different plant preferences but never utilize more than about 30% of the available plants. Some tropical nests may be 30 feet across and 8 to 9 feet in depth. Texas Leafcutter nests, although usually readily apparent due to the radiating trails and cluttered entrances, are rarely more than a few inches high. In my yard, *Atta texana* seem to like my fruit trees best of all, especially peach and pear, and will literally strip my trees if left unchecked. My Leafcutters are most active at night, but they also are out and about during the morning and evening hours and on overcast days.

Texas Leafcutters are fairly common in South Texas, and you are likely to find a colony in your yard or in the nearest park.

Geckos Are Unusually Common This Fall

OCTOBER 11, 1998

Our little native gecko—the Texas Banded Gecko—seems to be almost everywhere. It may be that the recent rains have brought them out of their normal hiding places, or perhaps it is just that time of year when lots of youngsters are out and about, or maybe it is a combination of both. But for whatever the reason, at least about my house, shed, and yard, they are more plentiful than usual.

They are most welcome, however, even those geckos that occasionally get inside the house. I know that they will help reduce the tiny bugs and spiders that one cannot help but bring in from the outside on clothing, as well as those that find their way inside in other ways. These little lizards will capture and consume almost any kind of tiny creature that they can find. They seem to have an enormous appetite, preying on spiders, roaches, flies, mosquitoes, and a variety of other arthropods.

Our Texas Banded Gecko often goes unnoticed because it is largely nocturnal, coming out of its hiding place only after dark (unless disturbed), but it can readily be found around outdoor lights and around windows where it hunts prey that is attracted to the lights. It also is our smallest reptile, only about 3 to 4 inches in length. Adults are banded with cream to yellowish bands against a chocolate brown to buff background. Their heads are mottled, and their underparts are grayish. Other characteristics include movable eyelids, unusual in lizards; four toes, each with a single claw; and a bulbous tail that is easily broken off. This latter feature undoubtedly aids them in escaping a predator that grabs their tail. Once detached, it will wriggle about on its own for several minutes, offering a choice morsel of food while the owner slips away to safety.

What's more, the Texas Banded Gecko is our only native lizard with a voice, a characteristic of all geckos. The name is derived from the interpretation of their call, "geck-o." The Texas Banded Gecko will emit its faint squeak when captured. But how they utilize this unusual ability in their normal existence is only conjectural. Probably it is part of their courtship.

We also have a nonnative gecko, the Mediterranean Gecko, in South Texas. It, too, is attracted to prey around lighted areas after dark. This larger gecko, about 4 to 5 inches in length, can easily be identified by its light brown color and warty appearance. It also possesses mouselike squeaks that can be heard at intervals on quiet nights. It apparently was accidentally introduced in the United States with shipments of materials from the Mediterranean region where it is native. It now occurs throughout the southern portion of the United States.

Five additional geckos occur in the States, but only one other—the Big Bend or Reticulated Gecko of the Texas Big Bend country—is native to Texas. Four others have been accidentally introduced into Florida: the Indo-Pacific Gecko is native to Southeast Asia, the East Indies, and many islands of the South Seas; the Ashy Gecko is native to Cuba, Hispaniola, and adjacent islands; the Ocellated Gecko is native to Jamaica and southern Caribbean Islands; and the Yellow-headed Gecko is native to central South America.

Our native Texas Banded Gecko seems to be doing very well, even with the competition from the Mediterranean Gecko. Although it is shy and will readily run away, it can be trained to feed on mealworms and flies when kept in captivity. It is one of our most abundant resident lizards.

Mesquite Beans Are Valuable Food for Wildlife and Humans

OCTOBER 12, 1997

Mesquite pods are fully ripe by September and drop in October. Sometimes the ground beneath the trees is literally covered with the 6- to 10-inch pods. The beans make choice foods for lots of wildlife, including deer, javelina, woodrats, and even some of our predators like coyotes and foxes. Not many people take advantage of these beans anymore, but there was a time when Native Americans and settlers in the Southwest considered them an extremely valuable natural resource. The U.S. Cavalry, while chasing Native Americans in West Texas and New Mexico, paid three cents a pound for mesquite beans.

Both the mesquite leaves and pods contain up to 13% protein and 36% sucrose, twice as much sugar as beets or sugarcane. The green pods can actually be chewed for their sugar content, they can be cooked into a mesquite syrup that is great on pancakes, and a highly intoxicating beverage can be made from the fresh sugary pods. Once matured, the beans can be ground into flour that is good for cornbread, pan bread, and cookies.

Getting to the beans inside the pods is no easy task, however. Step one is to toast the pods for a couple of hours at 150 to 200 degrees until they are brittle. Then break them up and run them through a grinder two or more times before sifting the material through a sieve, keeping the fine material as meal. Native Americans did the same thing with a grinding stone, called a *metate*, and by sifting the material through basketry made from reeds and willows. The resultant pan bread could be eaten then or stored for later use.

For those of you with an itch to try this, here is a recipe: use 1 cup of mesquite meal, 1 cup of whole wheat flour, and 1 cup of water. Simply combine the flours and water to make a dough, then pat out flat patties, and heat them in a skillet in a thin layer of oil; turn them over when brown. Serve them with butter and honey.

We take so much for granted today, buying whatever foods we desire at the local grocery, but early Texans could not run down to the store every time something caught their fancy. Not only did they have to plan ahead, but they had to spend every waking minute searching for or preparing for their next meal. They learned a great deal about what could and could not be eaten from the local Native Americans. They also learned a great deal about other uses of native plants and animals. But few native plants were as valuable as mesquite.

Native Americans used the mesquite's brown gum for dyes and paint and for mending pots. Medicinally, a tea made from the leaves and gum has been used as eyewash and for sore throats; tea made from the leaves and inner bark has been used as an emetic; and a boiled gum drink served as a purgative. Cradle boards were made from the roots, sharpened snags were used as plows, and the hard wood was used for hubs and spokes of wagon wheels.

More recently, mesquite firewood has become popular for barbecues due to its slow, intense burn and unique aroma. The plant is also used for carving, furniture, and fencing, and the flowering trees produce fragrant yellow flowers that honeybees utilize for producing a distinct, clear, amber-colored, and sought-after honey. All year long, mesquites provide an array of valuable products.

Impaling Prey Part of Butcher Bird's Prowess

OCTOBER 15, 1995

Loggerhead Shrikes are back in numbers. These little predators are again commonplace on fence and utility lines throughout South Texas.

Although a few individual shrikes remain in South Texas through the spring and summer months, the majority of our wintering birds migrate north in spring to nest elsewhere. But now, their harsh trill or rattle calls can be heard at almost any open field, and they are easily identified by their stocky, short-necked appearance, short wings, and black-and-white colors: black wings, tail, and mask; gray back; and white underparts and wing patch evident in flight. Its black mask makes it look like a little avian bandit. And it flies in a straight line with fast on-and-off wing beats.

The Loggerhead Shrike is most unusual in a number of ways. Unlike most other songbirds, it preys on birds, mammals, lizards, and small snakes, as well as a wide variety of insects, but because it does not possess sharp talons to tear its prey apart, as do the larger raptors (hawks, eagles, falcons, and owls); it must utilize the tools available. Therefore, it has learned to impale its prey on sharp yucca leaves, cactus, and other thorns, barbed wire, and such. It can then feed on the carcass for several days. It is not uncommon to find several prey species impaled on a fence or on a certain spiny shrub. The prey is almost always suspended with its head up and body hanging down. This impaling behavior has given the Loggerhead Shrike the name "butcher bird."

Recent studies have shed new light on the shrike's unusual behavior. On its nesting ground where impaled prey are most evident, the numerous impaled prey species represent the male shrike's hunting prowess in attracting a female shrike. Males with the larger number of impaled prey are first to attract a mate. Although both sexes impale prey year-round, decorated spiny structures are most common on their nesting grounds.

As many as 72 species of shrikes are known worldwide, but only 2 occur in North America, the Northern Shrike of the boreal forests and the Loggerhead Shrike of the central and southern states. All are small to medium-sized birds, 7 to 10 inches long, with large, broad heads and stout bills that are strongly hooked and notched at the tip. The notched bill is very similar to that of falcons. It includes a toothlike structure on the cutting edge of the upper mandible that corresponds to a notch on the lower mandible. These "teeth" are important in the shrike's killing ability. It is able to kill prey with a series of sharp bites with its strong hooked bill which can sever neck vertebrae of its prey.

Small prey can be swallowed whole, and their feathers and bones later regurgitated, but larger prey are carried to favorite sites and impaled where they can be eaten at their leisure.

Kingfishers Are Back! Winter Isn't Far Behind

OCTOBER 16, 1994

The return of Belted Kingfishers to South Texas is a sure sign that the hot summer is over. These wonderful birds normally spend their summers along more northern rivers and streams from Central Texas to Alaska but migrate south for the winter months.

A few individuals may remain all summer, and some even nest as far south as central Mexico. However, their abundance in our neighborhoods during the winter months provides a wonderful opportunity to enjoy one of nature's most unique creations close-up and personal. The Belted Kingfishers are never far away in the Golden Crescent area. They perch on wires or posts over roadside ditches or other wetlands or fly by with loud, rattle calls.

Kingfishers are among our easiest birds to identify. It is a fairly large bird (about a foot in length) with a noticeably large stocky bill and blue and white plumage; females also possess a rusty belly band. A closer examination reveals a finely banded tail, back flecked with white, and a tiny white spot in front of each eye.

By watching one of these active birds, you will soon understand why it is called "kingfisher." It spends most of its daylight hours foraging for food that may include fish of varying size, frogs, crayfish, crabs, and almost anything else that lives in water. But their method of fishing is what is most exciting. They physically dive on their prey headfirst, from a perch or hovering position up to 40 feet high, often becoming totally submerged, sometimes for several seconds. They will then literally fly out of the water with their prey either tightly grasped or stabbed with their sharp bill, carried to a favorite perch, beat senseless, flipped into the air, and swallowed headfirst.

At times, the prey may be so large that it is impossible to swallow whole. In such cases, the bird will simply remain still to allow its rapid digestion to consume its catch that slowly slips down its gullet. The undigested scales and bones are regurgitated as pellets.

Kingfishers nest in dirt banks along rivers, constructing tunnels as far as 15 feet deep and slightly angled upward. At the end of the tunnel, it constructs a 6- to 10-inch-deep nest chamber. Its bill serves as a digging tool, and it pushes the loose dirt out with its small but strong feet. The construction takes three days to three weeks, depending upon the type of soil. The chamber is then lined with grass, feathers, and materials from its pellets. After the young are fledged, the parents teach the youngsters the art of fishing by dropping dead meals into the water for retrieval; within ten days, the fledglings are catching their own prey.

Although the Belted Kingfisher is our most common kingfisher species, two other kingfishers occur in the Golden Crescent: the tiny Green Kingfisher is resident along the Guadalupe and San Antonio Rivers, and an occasional Ringed Kingfisher visits our area from further south. The larger Ringed Kingfisher sports an all rust belly and more massive bill and deeper rattle call. The sighting of a Ringed Kingfisher should be reported to the Golden Crescent Nature Club.

Albinism in Birds Is Rare

OCTOBER 19, 1997

The recent photograph of an albino hummingbird that appeared in the *Victoria Advocate* prompted me to check a number of pertinent references about this truly rare characteristic. Although albinism is possible in any species of bird, total albinism is extremely rare. In fact, bird banders in California found that only 17 of 30,000 birds (about ½ of 1%) that were handled over a ten-year period had any evidence of albinism, and only a minute part of those were considered total albinos.

John Terres, in his marvelous reference book, *The Audubon Society Encyclopedia of North American Birds*, states that as of April 1965, a total of 304 bird species, representing 1,847 individuals, had been reported with albinism; 54 families of North American birds were included. The two birds reported most often were the American Robin (8% of all records) and the House Sparrow (5.5% of all records). Other birds reported regularly with albinism included blackbirds, crows, and hawks. Red and yellow birds, such as cardinals, orioles, and goldfinches, were reported least.

Terres points out that there are four degrees of albinism. "Total" albinistic birds have a complete absence of melanin (dark coloring pigment) from the feathers, eyes, and skin. The hummingbird in the *Advocate* appeared to have black eyes, so it was not a total albino. Of the 1,847 reported birds with albinism, only 7% were total albinos. "Incomplete" albinos are those with pigment on at least one of the three features; these usually have totally white feathers but possess normal eye or skin color. "Imperfect" albinos have partially reduced pigments in all three features, and "partial" albinos, the most common form, possess some feathers without pigment while the eyes and skin color are normal. Although some individuals may show albinistic characteristics for their entire lifetime, the white marking may come and go in others. Bird banders report recapturing normal birds that had been partially albino in a previous molt.

Although there is no known cause of albinism, it appears to be genetic because certain areas of the country report far more cases of albinism than others. In researching for my book on the American Robin (University of Texas Press, 1999), I discovered that Portland, Oregon, had well-established annual newspaper reports of albino robins. A marked albinistic strain appears in the robins of that locality, producing one or more albinos every year.

Just as there are albinistic characteristics in birds, there also are melanistic birds, those with an excess of black pigment in their plumage, but this is even less common. Melanism has been reported for only 29 species of North American birds. One reason for this low number, perhaps, is that it is less obvious in species that regularly include dark phased birds. For example, dark forms of Rough-legged and Short-tailed Hawks are expected. Dark Red-tailed Hawks, those that summer in northwestern North America and often winter in central Texas, are known as Harlan Hawks; they once were considered a separate species, but these dark features are only environmental characteristics and cannot be considered the opposite of albinism.

One can't help but wonder how all-white or all-black birds manage to fit in with other normal members of the species—apparently, not very well. The literature contains numerous examples of complete or even partial albinos being harassed by their peers. Terres gives examples of such treatment in Barn Swallows and Red-winged Blackbirds, and in a study of South African penguins, "three freakish young in a nesting colony (one had an entirely black head, another a white head, the third was a complete albino) were friendless, shunned, and generally abused by their companions."

I couldn't help but wonder if the albino hummingbird, that we humans admire as an oddity, gets the same kind of treatment from other hummers.

Status of Our State Birds

OCTOBER 23, 1994

The spring 1994 issue of *American Birds*, a publication of the National Audubon Society, contained an article by Frank Graham, Jr., about the status of all the official state birds. Graham reported that over the past twenty-five years, based on annual Breeding Bird Surveys, these bird populations have declined in twenty-two states and are stable or increasing in another twenty-two states. Seven states do not have surveys to assess bird populations, and two states chose domestic chickens: Rhode Island Red in Rhode Island and Blue Hen in Delaware.

Reading the article, I immediately checked out the status of the Northern Mockingbird, Texas's official state bird, along with Arkansas, Mississippi, and Florida. Adopted by the Texas Legislature in 1927, the mockingbird was proclaimed "a fighter for the protection of his home, falling, if need be, in its defense, like any Texan." In spite of the mockingbird being so abundant in South Texas, survey data indicate that it has declined by about 20% throughout the state. It also declined in Arkansas, Mississippi, and Florida during the twenty-five-year survey period.

Other state birds experiencing declines included the Scissor-tailed Flycatcher in Oklahoma; Northern Cardinal in Virginia and North Carolina; Baltimore Oriole in Maryland; Wood Thrush in the District of Columbia; Hermit Thrush in Vermont; Carolina Wren in South Carolina; Western Meadowlark in the states of Montana, Nebraska, Oregon, and Wyoming; Lark Bunting in Colorado; Purple Finch in New Hampshire; American Goldfinch in Iowa; Northern (Yellow-shafted) Flicker in Alabama; Ring-necked Pheasant in South Dakota; and Greater Roadrunner in New Mexico.

Not all of the news is bad, however. A number of state birds showed increased populations. American Robin populations continued their rise in Connecticut, Wisconsin, and Michigan. The American Goldfinch increased in Washington State; Northern Cardinal in Ohio; Eastern Bluebird in Missouri and New York; Mountain Bluebird in Idaho and Nevada; Black-capped Chickadee in Massachusetts and Maine; Brown Thrasher in Georgia; Cactus Wren in Arizona; Ruffed Grouse in Pennsylvania; and California Quail in California.

The interesting thing about these ratings is the sharp turnaround in bluebirds. A very few years ago, these two bird species were experiencing a serious decline, blamed primarily on the indiscriminate use of DDT. Bluebirds feed directly on the insects targeted by the spraying of this and other pesticides. However, both the Eastern and Mountain Bluebirds have made an amazing recovery since DDT was outlawed, and thousands of concerned citizens and organizations have placed bluebird boxes at strategic locations all across the country.

This is one success story that affects us all!

American Kestrels Are Back for the Winter Months

OCTOBER 25, 1998

American Kestrels, the smallest of all our hawks and falcons, have returned to their wintering grounds in South Texas. Most of these little falcons, about the size of a Killdeer, will remain all winter, not departing for their northern nesting grounds until April or May. For now, almost every field has one or two of these colorful and personable falcons. Most can be found perched on wires or atop posts and other high points, where they can watch for prey. Others may be found flying overhead, often calling their very distinct "killy-killy" notes.

American Kestrels are among our most numerous wintertime raptors.

Once known as "sparrow hawks" because sparrows are sometimes used as prey, they now are properly called "American Kestrels." The kestrel is not a true hawk, such as the much larger, wide-winged buteos—Red-tailed, Swainson's, and White-tailed Hawks—but it is a true falcon of the genus *Falco,* closely related to the Peregrine, Prairie, and Aplomado Falcons. Each possesses long pointed wings bent back at the wrist and can reach great flight speeds when necessary. A coursing falcon flies fast and direct, usually without interruption. And all four of the above falcons also possess the typical falcon face pattern of a long, black wedge (sideburns) against otherwise whitish cheeks.

Although the larger falcons prey on species such as ducks and shorebirds, kestrels take much smaller prey, including small rodents, birds, snakes, bats, frogs, lizards, and a variety of insects. Crickets and grasshoppers are favorite food items. These usually are located from either a perch or by hovering in midair, but kestrels also are able to capture prey, including birds and bats, in midair. From a perch, prey capture is undertaken by a swift, direct pounce from above, with talons extended. Usually the prey is taken to a perch to eat immediately, but kestrels also are known to store extra food in a protected niche for a few days.

Many folks consider the American Kestrel the most appealing of all our raptors. Perhaps this is because of their small size, but it also may relate to their ability to adapt to a variety of environments. They seem to do very well in urban settings, so long as they can find food in a backyard or field. They may even nest in tree cavities close to our homes, usually where there is an abundance of House Sparrows. But in the wild, they utilize a variety of nesting sites, ranging from tree cavities or old crow and jay nests, such as those in the Texas Pineywoods, to rock ledges on isolated cliffs in the Chisos and Guadalupe Mountains of West Texas.

Like most raptors, females are somewhat larger than the males, 10 to 12 inches in length with a wingspan of 22 to 25 inches. The males possess the brightest plumage. One of our most colorful raptors, they sport a rufous red back and tail, with a black subterminal band and white tip, bluish wings, pale reddish underparts, and white, black, bluish, and reddish head. They also possess a pair of black eye-spots on their nape, thought to be protective coloration; the watching "eyes" may confuse some predators. Females are similar but are not so brightly colored and lack the blue-gray-colored wings.

American Kestrels occur throughout the Americas, from Alaska south to Tierra del Fuego, the southern tip of South America, and from California east to the West Indies. On many of the Caribbean Islands, they are best known as "killy-killy." Kestrels very rarely nest in the Golden Crescent but are abundant here during the winter months. Those individuals, like our many human "snow-birds," may have come south from anywhere to the north.

Wherever they have come from, they are most welcome. American Kestrels are a favorite winter visitor to many of us who enjoy birds and the wonderful avian diversity that is available in South Texas.

It's Been a Good Year for Acorns

OCTOBER 26, 1997

The heavy rains we experienced in recent weeks reminded me of Chicken Little's warning that the sky is falling. My 6-inch rain gauge, located in my backyard, filled up more than twice during the downpours, and for a brief period of time, my normal route into town was blocked by high water at the Guadalupe River bridge on SH 447.

But what made it all seem so out of the ordinary was the tremendous fall of acorns hitting the roof and deck. At one time, it sounded as if it were hail instead of acorns. A few days later, it took me several hours to clean the gutters, sweep the deck and driveway, and pick up an amazingly great number of acorns, along with the regular amount of leaves and debris, from the lawn.

It seems that 1997 is an especially good acorn year, with a yield that should fatten all of the acorn-eating critters that live in our woods. Many of these, such as deer, hogs, fox, coyotes, squirrels, turkeys, quail, doves, woodpeckers, waterfowl, and a number of the larger songbirds, relish the mast available from these oak fruits. Acorns, in fact, are one of nature's most valuable food items, not so much because they are a preferred food, but because they constitute a good and abundantly available staple. They are the staff of life for many wildlife species. To the north, their greatest value is in the critical winter season when other foods are scarce. In the warmer climes, acorns are just one of many valuable wildlife foods.

The southeastern United States has the largest number of oak species, with seven species being "most valuable" as acorn producers: Water, Willow, Live, Southern Red, Post, Black-jack, Swamp-chestnut, and Laurel. Paul Cox and Patty Leslie, in *Texas Trees: A Friendly Guide*, list twenty oak species for Texas. The oak that dropped its load on my house and deck is Live Oak (*Quercus virginiana*), the most common of the oaks in the Golden Crescent. Its acorns, about 1-inch long, are somewhat larger than average, and so the sound of hitting the deck is obvious.

Live Oak, also known as "Encino," is a large spreading evergreen tree that can reach 50 feet tall with a crown of 100 feet in diameter. The state champion Live Oak at Goose Island State Park is 44 feet tall with a trunk circumference of 422 inches and a crown spread of 89 feet. Its thick, shiny, dark green leaves are 1 to 3 inches long with smooth margins or toothed with tiny spines; these fall primarily in early spring when new growth appears. The male and female flowers are borne in separate catkins on the same tree in spring. Male catkins, 2 to 3½ inches long, are hairy with a yellow calyx; female catkins are 1 to 3 inches long with three red stigmas. The wood of Live Oaks is hard, tough, and fine grained but hard to work. It has been used for shipbuilding, hubs, and cogs.

Live Oaks are hardy trees that can live in an amazing variety of places. They seem to get along just fine close to the Gulf, where there is considerably salinity, and also do very well in the limestone and clay soils inland. Its range extends from Florida to Virginia and west to central Texas, an area that roughly corresponds to that of the Wild Turkey. Also, Live Oaks are relatively free of insect pests and diseases, although they harbor lots of insects and epiphytic plants, such as ball moss and old-man's-beard.

But of all their characteristics, and in spite of the occasional clutter, their most valuable asset is their acorns.

Fire Ants Are Everywhere after the Rains

OCTOBER 30, 1994

Fire ants often receive lots of attention following the heavy rains that Texas and other Gulf Coast states experience in mid-October.

The *Victoria Advocate* carried a front page photograph by Frank Tilley of a floating swarm of the Brazilian Red Fire Ants "seeking dry ground to rebuild their mounds." These "floating mounds consist of entire colonies that may easily carry hundreds or thousands of ants." Actually, a mature colony of Brazilian Fire Ants can contain more than 200,000 individuals.

Biologists generally agree that nonnative fire ants are "out of control" along the entire Gulf Coast, having already spread across 250 million acres from Georgia to Texas, since they were imported to this country in the late 1930s. Pesticide controls, which have to date cost us about $200 million, have been generally unsuccessful. In fact, some biologists believe that mass pesticide spraying contributes to their increase, "weakening many native ant species in the area, allowing the multi-queen colonies of the imported fire ant to revive and flourish," according to a recent article in *BioScience*.

This ant's amazing survival ability is its use of multiqueen colonies. Not only do they rebound faster, but their greater densities help them fight off other ant species competing for food. The imported Red Fire Ant may possess ten times more colonies in an area than any native ant.

The significant increases in Brazilian Fire Ants during the last fifty years have had serious effects on many of our native wildlife species. Twenty years ago, anyone turning over a rock in a South Texas pasture could find spiders, centipedes, millipedes, and a number of other native invertebrates; now, there often is little more than fire ants. Researchers tell us that in some areas up to 90% of all native ant populations and 40% of all native insect species have been killed off by these nonnative fire ants. And the decline of many grassland birds, such as various sparrows, meadowlarks, bobwhite, and our wintering Loggerhead Shrike, has been linked to the spread of fire ant colonies. Also, many biologists blame fire ants, along with habitat destruction, for the very serious decline of our prairie-chickens. Even white-tailed deer fawns have been killed by Brazilian Fire Ants. A fire ant grabs onto its prey with its mouth and attacks its victims with a stinger on its abdomen.

The latest idea for controlling the nonnative fire ant is to introduce a parasitic fly from Brazil that controls them in that country. These phorid flies lay their eggs on or into the ant's thorax or head; the larvae eat the tissue, pupate in the body or head, and mature inside the colony where it can begin a new life cycle.

Scientists are convinced that the phorid flies will not affect other ants, including the native fire ants, but additional study is continuing.

November

Dragonflies and Damselflies Are Amazing Little Creatures

NOVEMBER 1, 1998

Dragonflies and damselflies are some of our most fascinating insects. They remind me of hummingbirds, with their ability to hover and fly forward and backward. Unlike any other insects, they can move their wings independently. They can fly up to 60 mph and can lift up to fifteen times their own weight. What perhaps is most important, especially after our recent rains, their diet consists of mosquitoes and other types of flying insects. They possess a voracious appetite. One dragonfly in captivity was fed 40 horseflies in two hours. They are capable of eating their own weight in food every half hour.

About 450 kinds of dragonflies and damselflies occur in North America, and more than 100 of these can be found within the Golden Crescent. An Internet post by Richard Orr and Bob Honig lists 75 species of dragonflies and 32 damselflies for the Houston area. These include such catchy names for dragonflies as "Petaltails," "Clubtails," "Darners," "Spiketails," "Cruisers," "Emeralds," and "Skimmers." Damselfly names include "Broad-winged," "Spreadwings," "Threadtails," and "Pond Damsels."

Orr and Honig point out that no good book is currently available on the Odonata (dragonflies and damselflies) for Texas, but Sidney Dunkle's two books, *Dragonflies of the Florida Peninsula and the Bahamas* and *Damselflies of Florida, Bermuda and the Bahamas*, are useful for identifying the various species. There also is a major book in press on the dragonflies of North America, replete with color photographs, that should be available by late 2000. In the

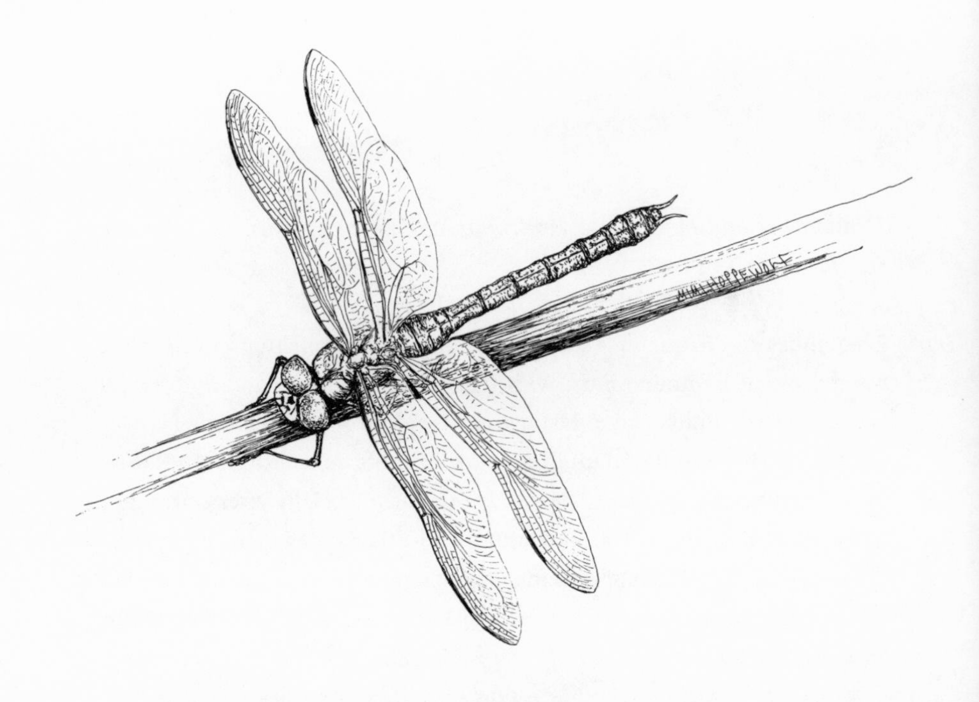

Dragonflies are extremely dexterous.

meantime, those of us with Internet access can get help from the Dragonfly Society of the Americas at www.afn.org/~iori/ or, for help with the identification of specimens, from Richard Orr at rorr@aphis.usda.gov.

In addition, Dr. Michael May at Rutgers University is gathering information on dragonfly migration during both spring and fall along the Gulf Coast and would like to hear from anyone who finds significant movements of these fascinating creatures. So, for those of you who might discover swarms of migrating dragonflies, you might want to contact him at mimay@rci.rutgers.edu or Dept. of Entomology, Rutgers University, New Brunswick, NJ 08903.

I suppose that it is inevitable that dragonflies and damselflies become the next great interest in natural history. Birds have long been at the forefront, and the increasing interest in butterflies has emerged in recent years, but suddenly all my birding friends are starting to get turned on by dragonflies. And I guess I have caught the bug, too, no pun intended. After all, with such neat names and their constant presence, it's easy.

Step one might be to recognize the difference between these two groups. Dragonflies rest with their wings held out, horizontally or nearly so, while damselflies rest with their wings held together near the body. Also, the hind wings of dragonflies are wider at the base than the front wings, while those of damselflies are similar in shape, both narrowed at the base. Male dragonflies possess three appendages at the end of the abdomen, while damselfly males possess four appendages.

Gloria Saylor of Port O'Connor told me that she has so far identified several of the dragonflies and damselflies found in her yard. Fairly common species include the Roseate and Golden-winged Skimmers, Eastern Pondhawk, Common Green and Swamp Darners, Blue Dasher, Wandering and Spot-winged Gliders, Common Whitetail, and Band-winged Dragonlet. She also reported the Great Pondhawk and Royal River Cruiser. How's that for some great-named Odonata?

Changing Leaf Color Is the Chemistry of Fall

NOVEMBER 3, 1996

The changing leaf color, from greens to reds, browns, and yellows, is commonplace all across the country, especially in the northern states, but even in South Texas our leaves change color, although less dramatically. The foliage of most of our broadleaf trees, and even cypress trees, turns from green to yellowish or brown, remains on the trees for a few to several more days, and then drops to the ground. In our yards, it is again time to rake.

Some eastern Native American tribes claimed that leaf changes were due to celestial hunters who killed the Great Bear and that his dripping blood fell onto the forest trees, gradually changing the leaves to various shades. And although "Jack Frost," or the actual occurrence of frost, has little to do with the changing colors, weather is involved. If the fall is rainy, cloudy, or very hot, the foliage generally becomes bland, yellowish, or less vivid. Sugars, which are manufactured by the leaves, are transported down into the trees where they have little effect on fall foliage.

Chemistry is most responsible for the color changes. Tree expert Robert Bartlett explained the process this way: "As summer wanes a band of tiny cells at the end of a leaf stem, where it hooks onto a twig, begin to dry and harden. This stops up the plumbing system inside the leaf. The manufacture of sugar slows down and the green chlorophyll no longer reaches the leaves. Now yellow pigments that have been masked within the leaves all summer are revealed. The red pigments are manufactured and the trees take on a kaleidoscope of hues and tones, a harmony of color."

Location and genetics also are significant factors in leaf colors. The southwestern side of a tree usually has the deepest color since it gets more sunshine. Trees in lower places may show color earlier than those in higher spots if cold air settles in the low spots on still nights and the cooler temperatures trap sugar earlier. Genetic differences are also important. Typical red leaves are found in maples, dogwoods, and Red and Scarlet Oaks. Browns and oranges are typical for White and Black Oaks, hickory, and hornbean, while yellows are more prominent in cottonwoods, pecan, redbud, and elm trees.

Even though our fall colors are less dramatic than they are to the north, they still represent a change in season, a time to appreciate the end of hot weather and the beginning of mild winter days.

Monarch Butterflies pass through South Texas in great numbers in the fall.

Monarch Butterflies Have Several Mimics

NOVEMBER 9, 1997

A lot has been written of late about the abundant Monarch Butterflies that are passing through South Texas this time of year. Hundreds or even thousands can sometimes be seen flying in a general southward direction, usually only a few feet above the ground, although others can often be seen much higher.

This mass movement too often results in great mortality along the roadways. Almost every vehicle inadvertently kills a few. Yet the masses continue their southbound migration, heading to mild climates in the mountains of central Mexico. Once there, they gather in great numbers, hanging from trees like fluttering draperies, and will remain there all winter. By early spring, they begin their northward journeys back through South Texas.

Few native predators bother the Monarchs because they are toxic and unpalatable. Monarchs acquire poisonous cardiac glycosides from the milky juices of their food plants, principally milkweeds. Although a few individual butterflies are attacked by predators, such as a bird or lizard, those predators rarely try a second time, and it seems that their hard-learned lesson also applies to a number of look-alike butterflies as well. Although the mimics may utilize different food plants and are not so toxic, they have developed a color and pattern very similar to that of a Monarch. It works!

Some of these mimics also fool butterfly watchers. Queens and Viceroys, and to a lesser extent Gulf Fritillaries and Variegated Fritillaries, possess many of the same features as Monarchs. All five species are orangish and black with whitish or silver spots. Monarchs are the largest, with a wingspan of 3½ to 4 inches, and orange wings with black veins and white-spotted black borders. The similar Queen Butterfly has a wingspan of 3 to 3½ inches, is browner and lacks the black veins but possesses large white spots on the tip of the forewings and in black borders. The Viceroy, less common in our area, is smaller yet (with a wingspan of 2½ to 3 inches), orange-brown with black veins, including a black line across the middle of the hindwings; both Monarchs and Queens lack the black line across the hindwings.

The fritillaries are very different when seen well. The Gulf Fritillary is a very common summertime butterfly, but it also flies during fall when the Monarchs are passing through. This is a beautiful creature with elongated wings (having a wingspan of 2½ to 3 inches) with bright orange upperparts (at least the male) with black margins, and underparts with large oblong, silver spots. Female Gulf Fritillaries appear faded. Variegated Fritillaries are somewhat smaller (wingspan of 2 to 2½ inches) and brownish orange wings with black markings, including a black spot within each of the outer cells; also they lack the silvery spots.

Last year was a very different butterfly year than we are so far experiencing in 1997. By September and October, multitudes of butterflies were flying, and it seemed like every flowering plant was busy with butterflies. At times, several other species, such as Snout Butterflies, could be seen by the thousands. And compared to the numbers of Gulf and Variegated Fritillaries nectaring in my yard, Monarchs were only a minor part of the equation.

This year, Monarchs seem to be our single most numerous fall butterfly.

Distinguishing the Ross's Goose Is No Small Task

NOVEMBER 10, 1996

Most folks admire the great flocks of Snow Geese that spend their winters in our area without ever noticing that as much as 5% of the flocks of the all white geese, with black wingtips and a reddish to pinkish bill, are actually a different species altogether.

This is the Ross's Goose, a smaller version of the Snow Goose that can be distinguished by its smaller size, stubby, triangular bill, and shorter neck;

the Ross's Goose call notes are higher pitched as well. Ross's Geese also lack the black "lips," a noticeable black patch between the upper and lower mandible, of adult Snow Geese.

Although an experienced birder can even distinguish the two species in flight, picking a Ross's Goose out of a flock of several hundred feeding Snow Geese in a field from 100 yards or more usually requires a good scope and considerable patience.

To make identification even more confusing, a small percentage of the Snow Geese flocks in our area include a darker bird with a dark gray body, bluish wings, and a white head and upper neck. It was once known as a Blue Goose, due to its bluish plumage, but we now know that this plumage simply represents a different color phase of Snow Geese. Yet the smaller but look-alike Ross's Goose is a different species.

Although the all-white and blue phases of Snow Geese nest together on the high Arctic tundra from northern Alaska east to Greenland, the Ross's Geese nesting range is limited. They utilize high Arctic tundra sites only from Canada's Northwest Territories east to Hudson Bay.

Wintering populations of Ross's Geese may have increased in the last few decades. When Roger Tory Peterson's first *Field Guide to the Birds of Texas* appeared in 1963, Ross's Goose was listed only as a "casual" winter visitor. Later field guides listed it occurring in "small numbers" along the Texas coast, but by the mid-1990s, it has become a fairly common winter resident. Maybe we have just learned better ways to identify it, or perhaps we are using better optics.

Whatever the reason, it occurs frequently but in small numbers wherever we find wintering flocks of Snow Geese. It is most welcome!

Flocking of Birds—Another Indication of Winter

NOVEMBER 12, 1995

The adage "birds of a feather flock together" is a truism that has withstood the test of time. In South Texas, huge flocks of blackbirds are commonplace during the fall, winter, and early spring months. The flocks are most apparent during the evening and early morning hours when they move between their roosting sites and feeding grounds.

At least six bird species make up these huge blackbird flocks, and each flock often mixes with two, three, or more species. Typically, flocks of Red-winged Blackbirds, Brown-headed Cowbirds, and Common and Great-tailed Grackles are most common, and a flock of any one of these may also contain a few of the others, as well as European Starlings and an occasional Bronzed Cowbird. At times the larger flocks appear like wisps of smoke or clouds in the distance. These birds often fly together, only inches apart in a synchronously tight flock like they are in formation, wheeling, diving, and ascending as one. They eventually will descend as one to alight in a field or pasture to feed on available seeds and insects.

Although blackbirds are best known for flocking in huge numbers, many other birds also flock, especially during the nonnesting season. Geese, ducks, cranes, Cattle Egrets, quail, swallows, waxwings, robins, chickadees, and even Wild Turkeys and cardinals flock. Although many bird flocks are simply extended families or a few associated family groups, several different songbird species often join mixed bird parties in winter. This is especially true in the Tropics. In Manu National Park in Amazonian Peru, as many as 70 species have been found in a single flock. They will spend the entire winter in preferred feeding areas. Closer to home, mixed flocks of 5 to 12 or more species often can be found in the company of a few full-time resident species, such as Carolina Chickadees and Tufted Titmice.

One can't help but wonder what advantage flocking might be for these birds. That question has interested ornithologists for a long time, and they have discovered a multitude of answers. Some are obvious. There undoubtedly is safety in numbers; at least one member of a flock is likely to detect a predator. It gives those birds feeding in the center of the flock more time to search for food and to eat in relative safety. New food sources can often be found, for the good of the whole flock, when the individual isn't spending the majority of its time watching out for predators.

Other answers are less obvious. Flocks flying in close quarters seem to fend off diving hawks, as a raptor will not dive into a solid flock for fear of injury. Flying flocks will bunch up whenever a predator appears. In the case of a mixed flock, composed of species with slightly different feeding patterns, they also are more likely to discover a greater variety of foods. Those feeding in the canopy may frighten insects into flight that are then captured by a bird feeding at a different level. And also in mixed flocks, some species tend to act as sentinels, others as guides, and others as beaters and searchers.

It seems that our wintering species, without the requirements of territorial defense, nesting, and feeding young, actually utilize a division of labor of sorts.

Why Do Woolly Bears Cross the Road?

NOVEMBER 15, 1998

It would be impossible not to answer such a question with the obvious reply, to get on the other side. But for woolly bears, the fat-bodied, hairy caterpillars of tiger moths, they are only looking for shelter to hibernate for the winter, usually under a rock or log, or in a secluded place to spin a rough cocoon. If their searching includes the crossing of roads, trails, driveways, yards and such, it is only part of their fall meandering. During their peregrinations, they have been clocked at up to 4 feet per minute, exceedingly swift for a caterpillar.

In South Texas, as well as throughout the country, we see more woolly bears in the fall than any other time of year. While the Northern Woolly Bears usually overwinter in the caterpillar stage, the Southern Woolly Bears (our South Texas species) will more often go into the next stage of their life cycle, a cocoon. As they spin, their body hairs are mixed with the silk, producing an oval brownish cocoon. This pupa stage will eventually become an adult tiger moth. Shortly after hatching, the females mate and lay egg clusters on various food plants. The eggs hatch in four or five days, and the caterpillars feed together on their host plants for three to four weeks, undergoing six molts before they are ready to pupate into the pupal stage.

For tiger moths, the woolly bear caterpillar stage is better known than the adults, but all are members of the Arctiidae family of moths, of which there are about 25,000 known species that occur worldwide, from the coldest regions to the hottest. Four species are fairly common in South Texas: Salt Marsh Moth, white with an orange yellow abdomen with black spots; Virginia Tiger Moth, with white wings flecked with a few faint black specks, and a yellow abdomen; Isabella Tiger Moth, with all yellowish wings flecked with tiny brown spots, and a reddish abdomen; and Leopard Tiger Moth, white with black squares, like a tiger.

Woolly bears are sometimes called "hedgehog caterpillars," due to their behavior of curling up into a tight ball when disturbed. This compact, hairy ball serves to ward off potential predators, like heavy armor or due to potential toxicity from the long spines. Such a ball also seems slick, easily sliding off the hand, so that it is difficult to grasp by predators as well as humans. Although the hair (actually known as *setae*) of some caterpillars is glandular, actually containing toxic fluids that can sting humans, that of woolly bears is not. Their purpose, that may serve to mimic the more toxic species, is more for the physical protection of the caterpillar against sudden changes in temperature, seldom necessary in South Texas.

In the northern states, some folks claim that woolly bears are weather forecasters that can predict the coming winter. They say that if the black bands at each end of the woolly bear are wider than the reddish brown central band, it will mean a harsh winter. However, if the central brownish band is much wider, the coming winter will be a mild one. Those in South Texas are all black, although an extended search will likely turn up a variety of banded individuals. The successes of such predictions are purely coincidental.

Winter Birds Are Looking to be Fed

NOVEMBER 16, 1997

Getting ready for the onslaught of winter birds? It is that time of year once again. So, if you plan on feeding birds, it is time to stock up with bird feed.

Although all of our wintering birds will do just fine without our help, feeding birds is great fun. It provides some close-up views of our feathered friends, and observing their behavior can be especially worthwhile. They will feed at close range and often seem unafraid. But to attract a whole range of winter birds, it is best to offer them what they prefer. And feeding birds also has a few obligations of maintaining feeders once started and making sure the feeders and the feeding sites are kept clean.

First the feed: to attract the greatest variety of birds, use a wide variety of feed. For seed-feeders, black-oil sunflower seed is best; striped sunflower seeds are bigger and have thicker seed coats, making them tough for small birds to handle and crack. Certain species prefer seeds other than sunflower.

For example, blackbirds like corn; doves are attracted to corn, milo, and millet; finches and goldfinches love thistle. Although many of the commercial "mixed bird feed" contains a blend of sunflower and other seeds such as milo, millet, oats, wheat, flax, and buckwheat, many birds will kick out the small seeds to get to the prize. Lots of seed is lost in this manner. The best bet is to use at least three feeders: one with a combination of feed (the mixed bird feed), one with black-oil sunflower seed, and another with thistle.

You also can attract insect-eating birds, such as chickadees, woodpeckers, nuthatches, wrens, and even warblers, to your yard by offering suet (beef fat) or peanut butter. A few of the seed-eaters, such as Chipping Sparrows, may also take a shine to these high energy foods. Suet or suet blocks, sold at most stores carrying bird feed, should be hung in a way that they do not attract raccoons; those omnivores can destroy your feeders overnight. Hang your suet in a wire basket or plastic mesh bag (the kind onions come in). Drill holes in a small log and press the peanut butter into the holes. It is best to mix the peanut butter with cornmeal; it cuts the cost of the peanut butter and makes it less gooey. Small birds have been seen choking on a mouthful of pure peanut butter.

Fruit also makes a choice food for a variety of birds, including robins, mockingbirds, and jays. Use a hanging tray or platform, or impale an orange, apple, or grapefruit half on a stake positioned on a hanging fruit feeder. Fruits, such as raisins and various dried fruits (soaked first), can also be placed on feeding trays.

Whatever seed you decide to provide, be sure to store it in tight, waterproof containers, and rake below your feeders on a regular basis; decomposing hulks can spread disease to your feeder birds. The feeders themselves should also be cleaned two or three times each year; scrub them with soap and water and dip them into a weak bleach/water solution. Be sure to rinse the feeders well and dry them before refilling them with seed.

One last thing, if you have been feeding hummingbirds, it is perfectly all right to continue during the winter months. Most of these marvelous little birds will have left for their wintering grounds to the south, but a few may chose to remain. They will continue using your feeders throughout the winter months without endangering themselves, even during very cold weather. But a word of warning: these wintering hummers depend upon a continuous supply of sugar water, so you must not stop the feeding until warmer weather.

If you feed birds, you're in good company. More than 65 million Americans provide food for wild birds, spending more than $2.6 billion on birdseed and bird-feeding supplies each year.

The Turkey Is Our Symbol of Thanksgiving

NOVEMBER 19, 1995

The turkey, that huge, ungainly bird of the oak woodlands and barnyards, has become a common symbol of Thanksgiving. No "Turkey Day" would be complete without it.

We are told that turkey is one of our most nutritious and healthy foods; we are encouraged to eat it year-round. For me, however, except for that marvelous smell from the oven, freshly cooked turkey is rarely as appealing as it is two or more days later. I like it best when I can pick the remaining meat from the bones or eat it in a sandwich or enchiladas.

The vast majority of the frozen turkeys we purchase at the local grocery stores are mass-produced and are a far cry from the critters that occur in the wild. Harry Oberholser, in *The Bird Life of Texas*, provides a marvelous comparison: "A wild gobbler has an alert eye in a slender blue head, a streamlined body covered with highly burnished feathers, and long legs; the domestic bird has a dull eye sunk into a swollen red head, a flabby body clothed in dirty feathers, and dumpy legs. The former bird runs better than a race horse through the woods and flies as lightly as a ruffed grouse; the latter can scarcely walk about its pen, much less fly." Wild Turkeys can be separated from feral turkeys by the tuftlike beard hanging from the wild male (occasionally female) bird's chest.

Wild Turkeys are magnificent birds, especially a courting male that struts about in a pompous manner with a fanlike tail display and expressive gobbling. The polygamous gobbler maintains a sizable harem, fending off rival males. The hen hides her nest with great care and protects it and a dozen or so eggs against predators and other invaders. Incubation takes about twenty-eight days, but the poults are so precocial that the hen and young leave the nest immediately after the last egg has hatched. Within three weeks, they are able to fly well enough to perch overnight in trees. By fall, when acorns are ripe, several families may congregate into huge feeding flocks.

The Native Americans had already domesticated Wild Turkeys by the time the first Europeans reached North America, and it was one of the very few animal imports to the Old World. The abundant Wild Turkeys of the New World (estimated at 10 million) became all important to the settlers. They were so important, in fact, that when our "national bird" was chosen, the turkey was second only to the Bald Eagle. Benjamin Franklin defended the turkey as his choice because it was "more respectable" than the thieving, scavenging Bald Eagle.

As Europeans moved westward, Wild Turkeys were plentiful throughout the eastern half of the continent, but by the late 1800s, only remnants of the original population remained. So, today, the Texas turkey population is largely a product of reintroductions from other localities, but because of its adaptable nature, it has become commonplace again in our fields and woods. No other creature is so representative of our natural world, as we prepare to give thanks for our many blessing, as our Wild Turkey.

Rails Are Some of Our Most Secretive Birds

NOVEMBER 22, 1998

Although most rails are rarely seen, even by the avid birder, six species overwinter in South Texas. Three kinds of rails are with us year-round. Perhaps the best known of these is the Clapper Rail, a foot-tall bird that is resident in saltwater marshes. Its loud "kek kek kek" notes are commonplace at dawn and dusk, and occasional individuals hunt food in the open on the mud flats near grasses, from where they can readily retreat when threatened. They can be recognized by their gray-brown upperparts and buff underparts with lightly barred flanks and long bill and legs. The very similar King Rail generally prefers freshwater marshes and possesses brighter plumage, including a cinnamon breast.

Our other full-time resident rail is the tiny Black Rail, only slightly larger than a sparrow and easily identified by its overall blackish color. One of our shyest of birds, it is usually detected first by its very distinct calls, a "kick-keer," that is often repeated incessantly in spring. It can be very aggressive, however, and also is known to utter wheezy chatter interspersed with hoarse rattles and vigorous "foot stamping" and hissing, utilized in defending a territory against an opponent.

During the fall months, three additional rails migrate from their more northern nesting grounds to South Texas, where they remain until spring. These include the fairly common Sora, uncommon Virginia Rail, and the very bashful Yellow Rail. Soras can be found in all the wetland habitats, from salt to fresh water. They are fat little birds, almost 9 inches in length, with brownish olive plumage, greenish legs, and a fairly short yellowish bill. Their call is a plaintive "ker-wee" or "wee-ker" and a descending whinny. Like Clappers and King Rails, Soras can often be found searching for food at the edges of marsh grasses.

Virginia Rails are never common in South Texas, but a thorough search in the proper habitat (both salt and freshwater wetlands, especially among cattails) can often turn up one or more of these elusive rails. Like most rails, they are most often detected first by their rather distinct call, an emphatic "kid kid kidick kidick." They also give a descending series of "oink oink" notes. These calls are rare in winter but sometimes given at dawn and dusk. Virginia Rails look at first glance like a small King Rail but can also be separated (at least the adults) by their size, gray cheeks, and reddish bill and legs.

Finally, the Yellow Rail is another of the tiny rails, about 7 inches in length with deep tawny yellow plumage above with dark stripes marked by white crossbars. It, like the Black Rail, is so shy that it seldom if ever is seen in the open or even by walking through the proper habitat that includes grassy wetlands as well as rice fields. Although they can be detected by their typical calls, four- or five-note "tick-tick, tick-tick-tick" (like the sound made by tapping two pebbles together), most observations are made by dragging a rope between two walkers through the grasslands. This is likely to flush the rails along the pathway, so they can be identified in flight. The Yellow Rail flight pattern shows a large white patch on the trailing edges of their wings; this is absent in the slightly smaller Black Rail and somewhat larger Sora.

One would think that rails, living most of their lives on the ground, would have a difficult time surviving the abundant predators, including alligators, snakes, and more, that live in the wetlands, but rails have learned to adapt to these various dangers by being extremely watchful, able to detect even the slightest motion, and by constructing mock nests before actually selecting one for raising a family. The disappearance of rails from our wetlands would not be due to natural causes but to human interference, especially the loss of essential habitat from developments and reclamation and from a variety of biocides. The survival of these enigmatic and mysterious creatures depends upon those of us who care.

Red-shouldered Hawk, Our Common Neighborhood Raptor

NOVEMBER 27, 1994

Sunday afternoon, while watching the Cowboys win again, I was attracted to a very different skirmish in my front yard. A Red-shouldered Hawk suddenly swooped to the ground and pinned down its prey. I immediately shifted my attention to the hawk and its captured critter. A few seconds later, it flew up into a Live Oak with a small snake that it held tightly in its talons. Within only a few more minutes, it had killed the snake, that I had by then identified it as a young whipsnake, and swallowed it whole. The hawk then flew off, apparently in search of another meal.

My Red-shouldered Hawk was a beautifully marked male with a chestnut barred breast and shoulders, blackish tail with three white bands, and back and wings flecked with white. It also had bright yellow legs and feet. Smaller and thinner than our Red-tailed Hawks, Red-shoulders are our only full-time resident woodland hawk; Red-tails prefer open country. In a sense, Red-shoulders are diurnal counterparts to the nocturnal Barred Owls, just as Red-tails are to Great Horned Owls of the open country.

Sometimes called "swamp hawks" because of their affinity for moist woodlands, Red-shouldered Hawks occur throughout the eastern half of the United States, from Florida to Quebec and west to central Texas and northern Tamaulipas; there also is an isolated population on the West Coast. Wherever they occur, their diet consists primarily of small native creatures. Of 220 stomachs studied by scientists, 102 contained mice; 40 held other mammals, such as moles, chipmunks, rabbits, muskrats, skunks, and opossums; 92 contained insects, especially beetles and grasshoppers; 39 contained frogs and toads; 20 snakes and lizards; 16 spiders; 7 crayfish; 3 fish; 3 poultry; and 1 earthworms.

Scientists also found that while studying 30 pairs of Red-shoulders, all remained within a 15-mile radius of their nest sites. One pair and its offspring occupied the same tract for forty-five years. Courting gets underway in February or March and is obvious by much calling and flying over their territory. They will circle low over their nest sites or may rise in ascending spirals to 1,500 or 2,000 feet, then dive toward their mate, all the while calling loud "kee-you" or "kee-aah" notes. When agitated, they may give a rasping "cac cac cac" call. They are the noisiest of all North American raptors.

Nests are well-built structures, normally on a large tree and sometimes constructed on top of an old nest of a Barred Owl or Fox Squirrel; there is one record of a Red-shouldered Hawk and a Barred Owl nesting in the same tree. Nests are built primarily of sticks, with a cup of mosses, lichens, leaves, and twigs. During incubation, the nests become well decorated with bits of white down from the hawk's breast. By April or May, fledglings are already out and about, crying for handouts and learning how to hunt and survive. Like their parents before, they will soon be part of our neighborhood.

Brewer's Blackbirds Are Often Ignored in Winter

NOVEMBER 29, 1998

Anyone driving the highways in South Texas during the winter months cannot help but find flocks of Brewer's Blackbirds feeding along the shoulders. These "winter Texans" arrive in October and November from their breeding grounds on the northern plains, and most will remain with us until March or April. They sometimes mix with other blackbirds, cowbirds, and grackles, and even European Starlings, but Brewer's Blackbirds seem to have a distinct preference for foraging along roadsides.

Brewer's Blackbirds can hardly be confused with any other bird because of their habitat preference and their general appearance and behavior. Males are coal black with a bit of a purplish hue on their head and yellowish eyes. Females are a drab brown color with black eyes. Both are relatively long-tailed birds, although not so long-tailed as Great-tailed Grackles (that also may occur along roadsides), and they walk rather than hop. Almost any very dark, mid-sized bird found walking is likely to be a Brewer's Blackbird. Red-winged Blackbirds also walk, but the male of these dark-eyed birds possesses a red epaulet (wing-patch), and they usually occur near wetlands.

Brewer's Blackbirds are one of the most gregarious birds. They, like Redwings, remain in flocks year-round. Most other blackbirds, such as cowbirds and grackles, flock in winter but occur only with their mates or in small family groups during the breeding season. The Brewer's Blackbird is colonial, nesting in colonies of a few to 100 pairs on sagebrush or on the ground in the more northwestern states. And even when nesting, they continue to congregate at feeding sites.

One can't help but wonder about the advantage of colonies. Such togetherness must afford them some benefits. Studies have proven the adage that there is safety in numbers. Out of many individuals, at least a few are likely to spot a predator and warn their neighbors, and individuals are able to spend more time searching for food when they are not constantly on watch. However, a huge flock of feeding birds is more likely to consume a food source before all the members feed. Yet their "rolling" foraging behavior covers considerable ground and seems to work very well. This rolling behavior consists of birds in the back of the flock flying over their flock-mates to the front until they again find themselves at the rear of the group. The rolling foraging style is very evident when watching a flock of Brewer's Blackbirds along a roadside.

Communal roosts are also an apparent advantage, for some of the same reasons. Although Brewer's Blackbirds are rarely found in the huge blackbird roosts in towns and cities, they do join blackbird roosts in the country, and some of those wintertime roosts can be astounding. Roost counts by the U.S. Fish and Wildlife Service in 1974–1975, principally in the southern states, included 723 major roosts with a total of 438 million blackbirds and 99 million starlings. In Texas, one roost had an estimated 50 million and another had 25 million birds. Of the 438 million blackbirds, 190 million were Redwings, 99 million starlings (not a blackbird), 91 million Brown-headed Cowbirds, and 10 million Brewer's Blackbirds. The remaining number included Rusty and Yellow-headed Blackbirds, grackles, and Bronzed Cowbirds.

These huge blackbird numbers require an amazing amount of food for their survival. This sometimes means flying considerable distances to find food. The energy expended by the individual bird can be considerable. The larger the bird, the more food required. Biologists at the University of Texas and Rice University, who monitored Great-tailed Grackle and cowbird numbers, claimed that the decline of males in late winter far exceeded declining numbers of smaller females. But it seems that Brewer's Blackbirds, that usually roost near their feeding sites and have a higher than average survival rate, may have developed the best strategy.

December

Perky "Butter-butts" Arrive for the Winter

DECEMBER 1, 1996

Early winter mornings in South Texas would not be the same without the perky little Yellow-rumped Warblers. Often known as "butter-butts" to birders, these energetic little warblers occur throughout our area in winter. They frequent our yards, woodlands, and field edges, and their lively behavior, sharp "chupt" or "tsup" notes, and bright yellow rumps add much to our wintertime bird life.

Once known as "myrtle warbler" because of their affinity for myrtle plants in the East, they were named Yellow-rumped Warbler in 1983 when they were lumped with their western cousins, the Audubon's Warbler. Both forms possess the bright yellow rump, although the eastern birds have a white throat and the western birds have a yellow throat.

Our wintering eastern Yellow-rumps feed principally on insects and other tiny arthropods, which they capture on vegetation or in midair like a flycatcher. However, unlike many of the warblers, this one has adapted to a winter diet consisting of seeds and fruits as well as arthropods, a major reason why this species is able to survive in colder climates. The only other warblers that we are likely to see in South Texas in winter are the Orange-crowned and Pine, although a few other species occasionally may remain as well.

Yellow-rumps remain in our area until April or early May, after which they move north to nesting grounds in the conifer and mixed forests of northern North America. Their only Texas nesting sites occur in the

The Yellow-rumped *Warbler* is our most common wintering warbler.

Guadalupe and Davis Mountain highlands, where the yellow-throated birds can be found. The white-throated birds nest much farther north along the Canadian border. The two forms overlap in the Canadian Rocky Mountains. They actually interbreed in these areas of overlap, the principal reason that the two forms, which can easily be separated in the field, were lumped together into a single species: butter-butt, or Yellow-rumped Warbler.

But whatever their affinity, they are most welcome to our neighborhoods.

Spiders Weave Nature's Own Christmas Tinsel

DECEMBER 10, 1995

Spiderwebs, those intricate patterns of spider silk, show up extra well on foggy mornings. Dozens can be seen among the oaks and other trees, each web dripping with condensation. The wet webbing gives them a silvery appearance, reminiscent of tinsel and other Christmas decorations. I recall reading somewhere that spider webbing gave the first Christmas decorators the idea of tinsel. It makes good sense.

Although late summer is a better time to find a variety of spiders, our recent foggy days highlight the last remaining webs. Before long, winter weather will kill the majority of adult spiders, and their webs will soon disappear, but until then, it is great fun to admire their handiwork—such intricate designs, and each species builds a different one.

The majority of the spider webs still hanging between the trees, usually from a couple to 15 feet high, were built by either the common garden spider or the Hump-backed or Spiny-bellied Orb Weavers.

Web construction varies with the species, but most spiders that build webs above ground utilize the same basic methods. Step one is a "bridge line" between two principle objects, that is pulled downward into a Y. The spiders then spin a second bridge line that is linked on all sides, forming a frame. Afterward, numerous strands are built between the frame and the center, and these are strengthened with circles of connecting webbing that complete the design. The last strands are sticky, so that they help capture prey. Many of these webs can be constructed, from start to finish, in sixty minutes. The final product is a true wonder of nature, more beautiful than any of our Christmas decorations.

What is also amazing is that spider silk is the strongest natural fiber known. Even steel drawn out to the same diameter is not as strong. The giant webs of some South Pacific spiders are actually used as fishing nets, and some spider webbing in Central and South America are used by native peoples as bandages. A scientist once stripped 6 feet of silk a minute from one of the tropical spiders, stopping after some 450 feet had been extracted; more was available.

Our natural world is truly remarkable. And as we begin to decorate our tree and home for Christmas, let's not forget the lessons of the world outdoors.

The Christmas Tree Is a Rich Tradition

DECEMBER 11, 1994

Christmas in many parts of the country is a snowman-and-skiing time of year, but for those of us in South Texas, where a white Christmas means little more than a nostalgic melody, the Christmas tree is our most obvious and cherished symbol of the season.

It is that time of year filled with greeting cards, Santa, "goodwill to man," and Christ. But whatever our religious preference, the Christmas tree seems to stand apart as being something special. A bright tree, covered with tinsel and bulbs, has a priority place in all our homes year after year.

No one knows for sure where the Christmas tree symbol began. Scandinavians once worshipped trees, and when they became Christians, evergreen trees became part of their Christian festivals. Others argue that it originated with Martin Luther, who, about 1500, tried to reproduce an outdoor scene of snow-covered pines, complete with the Star of Bethlehem, within his home. By 1561, an ordinance in Strasbourg, France, limited residents from cutting bushes for yuletide "more than the length of eight shoes."

Ornaments may have begun with precivilized humans who hung meat and other food on trees to keep it safe from wild animals. Those goods may have evolved into cookies and candies, and eventually tinsel was added. Do you remember the story of how tinsel was invented? It was after spiders made a mess of things, spinning webs over a poor woman's tree for her children, that a fairy godmother turned the trick to treat.

During the Great Depression in the 1930s, nurserymen could not sell their cultivated evergreens for landscaping and began to cut them for Christmas trees. Now, more than two-thirds of all Christmas trees sold in America are plantation-grown trees. Of 40 million trees grown and cut annually, 27% are Scotch Pine, 22% are Douglas Fir, 12% are Balsam Fir, and the remainder includes a wide variety of pine and fir. Balsam Fir is normally the most expensive because it usually possesses the perfect "Christmas tree shape" and retains its needles longest.

Elizabeth Silverthorne, in her lovely book, *Christmas in Texas*, points out that during the 1800s, after President Franklin Pierce first brought a Christmas tree into the White House, "Texans decorated their trees with whatever was handy: red berries, moss, mistletoe, cotton, pecans wrapped in colored cloth or paper, strings of popcorn, red peppers made into garlands, and homemade cookies and candies." She adds that "by the late 1800s Christmas trees were all the rage. The Austin *Statesman*' advised its readers: 'If you can't pay two dollars for one, take a hatchet, go out in the woods and poach on somebody's forest. You must have a Christmas tree or there will be no Christmas.'"

Attempting to explain the Christmas tree custom only fogs the fun. Trees will be part of the Christmas scene as long as kids from two to ninety get starlight in their eyes when they focus on the silver star atop the tree that, after all, symbolizes the real spirit of Christmas.

Mother Nature Decks out in Red Berries for the Holidays

DECEMBER 14, 1997

Bright red berries are commonplace at this time of year. They appear in late fall and early winter, seemingly to announce the upcoming holidays. But, in truth, it is all part of nature's plan to provide an added food source for the many birds that require the additional nutrients to help them make it through the colder and shorter days of the year.

Each year from late November to early December, our resident bird populations, including such stalwarts as mockingbirds, chickadees, titmice, and cardinals, will be joined by many other species from the north, including a variety of sparrows, a few warblers, and such wintering birds as Eastern

Phoebe, Northern Flicker, and American Robin, that will remain through the winter months.

A survey of the berry crop in yards and along the roadsides offers a surprising variety of shrubs and trees, and even a few vines, with red berries. Undoubtedly the most common, and also the most obvious of these, is the native yaupon (pronounced "YO-pahn") or Yaupon Holly. This is a shrub that occurs in brushy areas and has evergreen, lightly scalloped leaves and an abundance of red berries. It produces clusters of white flowers in spring and is filled with berries growing near the stems in fall and winter. A member of the holly family, it is a favorite of birds, and it also was utilized by Native Americans and early settlers. According to Paul Cox and Patty Leslie, in *Texas Trees: A Friendly Guide*, the "leaves contain a small amount of caffeine, and a coffee substitute can be made by steeping the dried leaves in hot water. Indians made a beverage called the 'Black Drink' which they used for a ritualistic purging ceremony—hence the species name *vomitoria*. Like other hollies, yaupon berries contain a poisonous substance that can cause vomiting and diarrhea."

Another native holly that produces red berries that mature in fall and remain on the shrubs through the winter is Winterberry or Possum Haw. It looks at first glance like yaupon, but its red berries are not so clumped and concentrated around the stems. It also prefers thicket habitats. And one of our most attractive shrubs is American Beautyberry, also uncommon in the understory of our woodlands. This native shrub produces lilac flowers in early summer and bright lavender berries that appear in masses around the pairs of thin, ovate leaves.

In addition to the native plant species, another six to seven nonnative shrubs with red berries can be found in our area: American, English, and Japanese Hollies; Cotoneaster; Sacred Bamboo; Tartar Honeysuckle; and Pyracantha. The most common of these is Pyracantha, or Firethorn, an evergreen shrub with small leathery, green leaves that produces multitudes of fleshy red berries in late fall. These usually remain on the plant all winter, or until they are consumed by the wintertime fruit-eaters. Although all berries can sour and ferment while still in place, Pyracantha berries are best known for this, and feeding birds can readily become intoxicated.

There are lots of records of robins and Cedar Waxwings becoming drunk after eating too many fermented berries. I have witnessed a half-dozen robins flopping about on the grass after eating fermented Pyracantha berries. They lie on their sides trying to keep their balance with outstretched wings, then, when trying to run, go in circles and eventually fall forward, smashing their bills into the ground. Some sit back on their tails with wings outstretched, but eventually fall over sidewise. Very drunk robins! Although they remained in this condition for several hours, they all eventually recovered to fly off in search of more bed berries.

Robins and other wintering birds concentrate in areas where there are plenty of berries and usually move on to other sites when they have consumed all the available berry crop. In good years, as this one seems to be, wintering berry-eaters can be abundant.

The Bright Red Cardinal Is a Showy Symbol of Christmas

DECEMBER 15, 1996

I can think of no other bird that so signifies the Christmas season as our bright red Northern Cardinal. It is destined to appear in nature magazines and a host of other types of magazines over the holidays. Some of these photographs and paintings are in natural surroundings, sometimes even with snow and icicles, but other times they appear in artificial settings among apples and oranges or even on a Christmas tree amid a group of other birds. Even the partridge, the one that is found in pear trees during the twelve days of Christmas, takes second place to our lovely "redbird."

There are, of course, good reasons for this extra exposure during the Christmas holidays. Principally, its cardinal-red color seems to be part of the season, but its overall appearance and personality are also fitting. Henry Nehrling, an author in the late 1800s, had this to say: "The cardinal is one of the jewels of our bird-fauna, being incomparable in the combination of proud bearing and gaudy coloring, and unexcelled in certain qualities of its song. Few birds impart their haunts with such life, beauty, and poetry as this brilliant songster, one of the most famous among birds and highly prized by all bird lovers."

As might be expected with one of our best known and most beloved birds, it has acquired a variety of names. Among those are "redbird," "big red," "cardinal bird," "topknot redbird," "cardinal grosbeak," and, in Vir-

Northern Cardinal males are a favorite Christmastime symbol.

ginia, even "Virginia nightingale." The "cardinal" name comes from the Latin word *cardo,* meaning "the hinge of a door," referring to the "importance that an idea hinged or depended," according to June Osborne, in her lovely little book, *The Cardinal,* published as a gift book by the University of Texas Press. June continues that, "Carollus Linnaeus, the famous eighteenth-century Swedish botanist known as the Father of Taxonomy, chose to ascribe the name 'cardinal' to the bird whose plumage matches the radiant color of the papal robes of the church's cardinal. Through the centuries the name has stuck."

Our cardinal has a rather extensive range from New Brunswick in the northeast, southwest across the southern half of North America to the edge of California, and southward throughout Mexico to Belize and Guatemala. It is nonmigratory, a year-round resident throughout its range, able to adapt to a wide range of habitats and weather conditions, truly an exceptionally hardy creature. Its principal enemy today is the house cat, one of the most dangerous of all predators!

Throughout its very extensive range, South Texas seems to contain as many or more cardinals than practically anywhere else. June points out that, according to Christmas Bird Counts that are undertaken annually throughout North America, "the densest concentrations of cardinals in winter occur on the Mississippi River, both in the South and farther north, and also along the Colorado and Guadalupe rivers in southern Texas. Less dense cardinal populations are found in winter along the Ohio, Arkansas, Brazos, and Red rivers."

But wherever they occur, they seem to be most evident around Christmastime. Perhaps that is when natural foods are most scarce, and they are spending more time at our feeders, but perhaps, it is because we are spending more time at home enjoying the holidays and are able to appreciate the comings and goings in the yard. The bright red male cardinal will certainly attract our attention!

'Tis Christmas Bird Count Time Again

DECEMBER 18, 1994

More than 44,000 volunteers will participate in about 1,700 Christmas Bird Counts from Alaska to Argentina this year, between December 16 and January 2. These annual CBCs represent the single largest wildlife census on record; participants are expected to find and identify almost 600 bird species in North America this year. Last year, participants tallied more than 60 million birds, and 70 of the 1,700 counts produced 150 species or more.

Of the 70 high counts, 17 were conducted in Texas, including three of the top ten: Corpus Christi led the nation with a grand total of 217 species; Mad Island Marsh was second with 205 species, and Freeport tied Moss Landing, California, for third place with 204 species. Other high Texas CBCs included Coastal Tip, Bolivar Peninsula, San Bernard National Wildlife Refuge (NWR), Corpus Christi (Flour Bluff), Aransas NWR, Port Aransas, Choke Canyon, Rockport, Attwater Prairie Chicken NWR, La Sal Vieja, Galveston, San Antonio, and Anzaldous-Bentsen. In 1994, 82 CBCs were conducted in Texas, second in number only to California's 110 counts.

Each CBC is limited to one twenty-four-hour period within the approximately two-week time period and within a designated 15-mile-diameter circle. Although preliminary scouting usually occurs, only birds found on count-day can be included in the final census. Over the years, these counts provide one of the very best indicators of bird populations and trends. Some of the very first North American CBCs began at the turn of the century, and those areas will soon have annual winter bird records for one hundred consecutive years. Victorians have participated in annual CBCs since 1977. In 1994, twenty participants tallied 43,236 individuals of 143 species on the Victoria Count.

But in spite of the great value of the CBCs, to many of us, Christmas Bird

Counts are as much a part of the holiday as trees and presents. They offer an additional excuse to be outdoors, enjoying nature and our natural heritage. For those individuals not able to get out, birds at feeders located within the count circle can also be counted. In fact, bird feeders often produce birds that are not found anywhere else. People with hummingbirds at their feeders are especially encouraged to participate, or at least to notify the count coordinator so that their birds can be included.

Experts and beginners alike are welcome to participate. Everyone is eligible to partake of the birder's Christmas spirit.

Flowering Poinsettias Are a Lovely Symbol of Christmas

DECEMBER 20, 1998

Almost everyone admires the bright red poinsettias so common during the Christmas season. These bright red flowers, that can be purchased in dozens of stores, brighten up any household, even among the already abundant decorations on the Christmas tree, fireplace, and elsewhere. In most households, Christmas would not be Christmas without one or more of these spectacular plants.

The Christmas Poinsettia is not native to the United States, however, but occurs in the wild only in southern Mexico and Central America. Known to scientists as *Euphorbia pulcherrima*, it came to Texas via South Carolina, where U.S. diplomat Joel R. Poinsett sold cuttings from Mexico to a nursery in 1828. The flower was named in his honor. Today, the plant is so popular that there is a national organization, the American Poinsettia Society, with headquarters at Mission, Texas.

In Mexico, poinsettias are known as Christmas Eve or Mexican Flame Leaf, and the Aztecs called it "cuitlaxochitz," meaning false flower. That term was derived from the characteristic flowerlike arrangements that are not flowers at all, but a series of highly colored bracts or modified leaves. The tiny, nondescript flowers, with yellowish anthers, appear in the center of the bright red bracts.

The Christmas Poinsettia is but one of more than 1,000 species of *Euphorbia* worldwide, and 62 species are known for Texas. Although they all possess similar characteristics, only the nonnative Christmas Poinsettia has the large, bright red bracts. All the Euphorbs also produce a milky sap that can be quite caustic; wear gloves and protect your eyes and mouth while pruning. Although these plants can be maintained in outdoor gardens, they are unlikely to survive a hard freeze. They normally are propagated by the millions for Christmastime sales, but they can be nurtured at home to produce their gorgeous "flowers" each Christmas.

If interested, as soon as the bracts fade in spring, cut the stems back to about 4 inches above the soil and replant them in a standard soil or peat-based compost. Or stem cuttings can be taken in spring; dip the bases in crushed charcoal to stop bleeding. Feed your plants every two weeks. About twelve weeks before you want your plants to produce the red bracts, place them where they will experience complete darkness each night. Your plants should be in full color in about twelve weeks.

As might be imagined, several legends surround our Christmas Poinsettia. One favorite legend is a Christmas tale of faith that Elizabeth Silverthorne includes in her lovely book, *Legends & Lore of Texas Wildflowers:*

> Many years after the birth of the Christ Child, a small ragged child stood weeping outside the walls of a great cathedral because she had no gift for the Christ Child and could not join the joyous procession marching to His altar. Suddenly an angel appeared and told her to give whatever she could and to believe that in the giving the gift would become lovely. Finding nothing in the churchyard but a thick, coarse weed, she broke it off and carried it with faith in her heart to the altar, where she placed it reverently. As she knelt there the tips of the plant suddenly blazed out into glorious red stars, and the watchers, who believed it represented faith and love made beautiful, called the flower *Flor de Navidad* (flower of the nativity) and *Flor de la Noche Buena* (flower of the holy night). . . . In Christian symbolism the red Christmas poinsettia came to stand for the Virgin wife, the blood of Christ and everlasting life.

A Symbol of Christmas—Mistletoe

DECEMBER 25, 1994

Christmas is the time of year that seems to pop up just when one is finally getting the Thanksgiving dinner fully digested. It is a Christian holiday that is so commercialized today that the original intent is all but lost to many Americans. And yet, two native symbols of the Christmas season have lasted over the years—the Christmas tree and mistletoe.

Mistletoe is full of life in winter when it seems that life for many plants is at its lowest ebb. Once gathered as a symbol of life and purity by the Druids of ancient Gaul, the mistletoe figures in legends of Germany and Scandinavia, and today is hung at Christmas as a promise of life and fertility. In many countries, a person caught standing beneath mistletoe must forfeit a kiss.

The plant belongs to the mistletoe family, Loranthaceae, which contains about 500 species that occur on a wide variety of woody plants throughout the tropical and neotropical regions of the world. Partly parasitic, it derives part of its nourishment from its host plant. The rest of its food is manufactured from the chlorophyll of its greenish yellow, leathery leaves. Tropical species may flower and fruit year-round, but more northern mistletoe plants flower in the spring and produce semitransparent berries in the fall and winter; many are at their peak at about Christmastime. The fruits are eaten by birds that often spread the plant by wiping the sticky glutinous seeds on branches of trees.

For years, people regarded the waxen berry as a charm against epilepsy, nightmares, and witchcraft. It has been considered a good luck piece in many parts of the world, worn in the lapel or around the neck to keep diseases away, placed under the pillows to induce dreams or omens, laid upon the threshold to prevent nightmares, carried by women to cure infertility, and placed in fields to stimulate crop fertility.

Mistletoe was once forbidden in Christian churches because it was thought tainted with heathenism, but now it is a symbol of life, along with wintergreens, and is brought into households at Christmastime as a decoration and also to perpetuate the pleasant custom of kissing.

South Texas Is a Winter Home for a Variety of Raptors

DECEMBER 28, 1997

The coastal prairies, pastures, resting croplands, and old fields offer lots of appealing habitat for a variety of northern raptors that move south for the winter months. Eagles, harriers, hawks, falcons, and even a few owls can usually be found with a little insight and a few hours in the field. The wintering raptors frequent the same areas utilized by our resident Red-tailed, Red-shouldered, and White-tailed Hawks and Barn, Barred, and Great Horned Owls, and Eastern Screech-Owls. For anyone with a desire to observe these fascinating creatures, the following suggestions will help you find some of these birds.

Peregrine Falcons are one of our most exciting raptors, able to dive at more than 100 mph and prey on anything from ducks to swallows in midair. Peregrines are most numerous along the coast and at wetlands. Although they can possibly be found almost anywhere in our area, a sure thing is in the Magnolia Beach–Indianola area. Such coastal sites also will produce Northern Harriers, White-tailed and Red-tailed Hawks, and American Kestrels, and watch, too, for Merlins, a falcon that nests in the northern forests. Great Horned Owls occur in this habitat year-round, and Short-eared Owls are present in winter.

Bald Eagles prefer wooded sites, especially along rivers or reservoirs. A pair has wintered and even nested in the Victoria Dupont Plant for several years, and Dupont has even constructed a viewing platform along their north entrance road from where visitors can view soaring birds. Other pairs can sometimes be seen along Coletoville Reservoir and the San Antonio River bottomlands, such as north of SH 239 west of Tivoli. These wooded areas also harbor Sharp-shinned and Cooper's Hawks (hawks that feed on a variety of small songbirds) and Red-shouldered Hawks. Barred Owls and Eastern Screech-Owls reside in this habitat as well. Victoria's Riverside Park is another good place to find Sharp-shinned, Cooper's, and Red-shouldered Hawks and the two owls.

American Kestrels, earlier known as sparrow hawks, are commonplace in winter in our abundant fields and pastures. This is a lovely little falcon with pointed wings and a black-and-white facial pattern; males possess a reddish back. It also is one of our most vocal hawks, calling distinct "killy-killy-killy" notes anytime and often. Salem Road, north of the Victoria city limits, is one of the better areas to find this and several other raptors of open areas. Red-tailed Hawks, Northern Harriers, and Crested Caracaras are usually common; White-tailed Hawks are less numerous; watch for White-tailed Kites (earlier called Black-shouldered Kites); and Ferruginous Hawks are present some winters. This large hawk looks at first glance like a Red-tailed without the reddish tail. Their tail and breast are whitish, and they may show chestnut thighs and a dark crescent on their underwings.

Also watch in the fields for Short-eared Owls and the less numerous Burrowing Owls. The larger Short-eared Owl is most active at dawn and dusk, flying low over the fields with a typical butterfly-like flight. Flying birds also show dark patches on their wingtips and wrists. The little Burrowing Owl is diurnal and overwinters in piles of concrete or other debris which can be used as burrows. It usually stays on or near the ground.

Finally, Barn Owls are resident year-round but are seldom common. These monkey-faced owls utilize old, usually deserted dwellings, such as old barns and houses. They possess whitish (actually a soft or golden yellow) plumage, with long legs and a heart-shaped facial disk and dark eyes. Almost the size of a Red-tailed Hawk, Barn Owls are most often detected by their very distinct calls, a drawn-out, raspy screech that grows louder and harsher toward the end. They also give a short, harsh hiss.

All of these raptors are present in winter. Each offers exciting observations and insight. Each is worthy of our care and protection!

Texas Coyotes: The Most Loved or Hated of Our Wildlife

DECEMBER 29, 1996

None of our wild animals are so loved by some and hated by others as our native coyote. They have long had a reputation for preying on smaller and weaker creatures and have raised considerable havoc in our yards, pastures, and even in our towns. Chickens, calves, sheep, and even pet dogs and cats have fallen prey to these wily critters. But for all of their negative habits, coyotes are one of the most enduring and adaptable creatures that ever existed.

This time of year is when they are most noticeable. From January until late spring is when you are most likely to hear their sorrowful songs. Their breeding season occurs during this period and so they are most active. Their loud wailings, strange barks, and "yip-yap" howls can be heard at dawn and dusk throughout South Texas. In my opinion there is nothing so satisfying as a dawn chorus of coyotes.

Coyotes must be one of our smartest wild creatures. They have survived in spite of extensive persecution by ranchers and others over the years. Although many individuals are annually killed by guns, traps, and poisons, they continue to increase, and they now reside all across the country, even in and adjacent to some of our larger eastern cities. Their survival proves an ancient ecological fact that stressed animals tend to reproduce at the highest level.

Almost everyone who loves nature must at least admire the coyote. Even those who hunt down Señor Coyote can't help but admire his cunning and curiosity. I once had a coyote follow me for more than a mile while I was walking a bird population transect. He stopped each time I did, actually sitting down and watching my every movement, and got up and followed me each time I moved on. On a couple occasions, he was within 50 feet or so, once he realized, I suppose, that I was not a threat. It was a marvelous experience, like being a real part of nature.

Texas author J. Frank Dobie also admired coyotes, as he wrote an entire book about them: *The Voice of the Coyote.* Dobie includes dozens of stories about coyotes, and it is a wonderful read. He tells about how a coyote led desert travelers to water, marvelous stories about their hunting prowess, including taking porcupines, their methods of group hunting, how they have outsmarted human trackers, and their unbelievable adaptive abilities. Anyone reading Dobie's book cannot help but admire Señor Coyote even more.

A New Year Brings New Hope

DECEMBER 31, 1995

What will the new year bring? Perhaps all our hopes and dreams will be fulfilled. Perhaps we will win the lottery or inherit a few million dollars from some long-lost ancestor. We can then do all the wonderful things that we have wanted to do for so many years.

Don't bet on it! The chance of becoming a multimillionaire overnight is so small that dwelling on such an occurrence would undoubtedly be counterproductive. It is best to stay with our day jobs, live a reasonably clean life, and save enough to guarantee a reasonably comfortable retirement.

But what if? I suppose that I would do a little more traveling to some exotic countries that possess huge resident bird numbers, maybe update my stereo and computer equipment to the state of the art, and give great hunks of money to various conservation organizations earmarked for specific projects.

There is so much that needs to be done right in our own backyard that I suspect my donations would wear pretty thin before I considered anything outside of Texas. For instance, acquisition of properties in West Texas, particularly in the Davis Mountains and outside Big Bend National Park, would take a huge chunk, and dozens of projects along the Gulf Coast and in the Lower Rio Grande Valley need help.

I suppose that my first step would be to prioritize my project list. The acquisition and/or protection of uncontaminated wetlands, especially freshwater wetlands, must be of highest priority. These types of areas, especially those few remaining ones along the Gulf Coast, urgently need our help.

In the Rio Grande Valley, an area that is fast becoming one long row of concrete, a few sites still deserve our attention. The Lower Rio Grande Valley "Wildlife Corridor," stretching from Falcon Dam to the Gulf, richly deserves our contributions.

Closer to home, maybe I would fund added road patrols to control roadside dumping and greater efforts to control feral house cats and dogs, and also establish an extensive environmental education program in our schools. After all, if our youngsters grow up with the same attitudes as their parents, our natural resources will continue to decline. Come to think of it, maybe that should be step one.

I guess I need to rethink my entire list of priorities.

Index